On Privacy-Preserving Protocols for Smart Metering Systems

Fábio Borges de Oliveira

On Privacy-Preserving Protocols for Smart Metering Systems

Security and Privacy in Smart Grids

 Springer

Fábio Borges de Oliveira
Laboratório Nacional de Computação
  Científica (LNCC) – Petrópolis
Rio de Janeiro, Brazil

ISBN 978-3-319-40717-3   ISBN 978-3-319-40718-0   (eBook)
DOI 10.1007/978-3-319-40718-0

Library of Congress Control Number: 2016943139

Printed on acid-free paper

This Springer imprint is published by Springer Nature
The registered company is Springer International Publishing AG Switzerland

To my parents,
even though they educated me and never bought me a Lab!

To my dissertation advisor,
even though he censured the above dedication in my Ph.D. thesis!

To my teachers,
even though they never answered all my whys!

To my professors,
even though they asked me too many whys!

Preface

Scope

The global economy and sustainability issues are driving suppliers to new operating modes. Smart grids and their smart metering systems can yield sustainable and profitable operating modes. Thus, smart grids are important enablers of economic development. However, along with benefits, smart grids bring drawbacks. Similar to other interconnected technologies, security and privacy are crucial to smart grids. Neglecting security concerns might eventually compromise, for instance, the supply of electricity, water, or gas. Neglecting privacy concerns might cause the violation of the right to privacy of customers, enable surveillance, and permit manipulation of all customers. Indeed, smart meters are becoming ubiquitous, and smart grids face unprecedented threats. Public infrastructures might be jeopardized, and citizens might be manipulated. Luckily, Privacy-Preserving Protocols (PPPs) can solve this impasse.

Audience

This book is intended to researchers and students who are interested in the area of security and privacy. Indeed, this book's focus is smart grids. However, the techniques presented in this book can be used in several application areas, for instance, electronic voting, reputation systems, sensor networks, electronic money, mobile sensing, multiparty computation, and image processing.

Recommended Background

This book covers the topics necessary to understand the bases of security and privacy. However, the background required to have a better understanding of some equations in Chap. 6 is knowledge of cryptography and an undergraduate degree in computer science or mathematics.

Content

This work advances state-of-the-art PPPs with the development of several protocols that preserve customers' privacy and security in smart grid scenarios. Four of them are revisited and improved in this book. Such development culminated in the concept of Asymmetric DC-Nets (ADC-Nets)—from "Dining Cryptographers"— which are generalizations of additive homomorphic encryption primitives. In addition, we can use such primitives to construct ADC-Nets, which are crypto-graphic primitives for encryption, aggregation, and decryption of aggregated data. ADC-Nets underlie secure, verifiable, efficient, and scalable protocols with low communication overhead, which are independent of trusted parties and resistant to collusion. Furthermore, smart meters can send the minimum number of required messages directly to their supplier. Thus, they can sign their messages, and as consequence, the protocols can ensure non-repudiation and fault tolerance. The former ensures that customers cannot deny the messages of their smart meters if these were transmitted. The latter ensures that their supplier can detect smart meters with failure—in themselves or in the communication channel—and can run the protocols without the compromised smart meters. Moreover, ADC-Nets can enforce customers' privacy.

Besides the concept and results of ADC-Nets, this book presents other contributions listed as follows:

- This book contextualizes smart metering systems in smart grids around the world and points out the needed models to have security and privacy in smart grid scenarios. Furthermore, it reviews the state of the art of privacy-enhancing technologies for smart metering systems.
- This book presents three scenarios, which require remote and frequent measurements. In addition, it assesses the minimum requirements for PPPs. Moreover, it is shown how computations can be done over encrypted measurements.
- An algebraic and a probabilistic analysis show that PPPs cannot keep customers' privacy secure using data aggregation with a small number of customers. Counterintuitively, when the number of measurements increases, the effectiveness of PPPs also increases. The optimal effectiveness is achieved when the sum of measurements and the number of smart meters are equal. These results are independent of PPPs.

- The four selected PPPs have different interesting properties. The first protocol leads to the conjecture that it has the fastest encryption algorithm, because it requires only a "one-way function." The second is based on elliptic curves, and further, the encryption algorithm uses only two scalar multiplications that lead to a fast protocol. The third uses an ADC-Net and inherits its benefits. When the level of security is increased, the second and the third protocol become increasingly faster than typical solutions. The fourth follows the laws of quantum mechanics, which surprisingly implies that the smart meters do not need to store a key, but they can send messages directly to their supplier without compromising privacy.
- To compare the protocols' performance, this book presents simulations with millions of real-world measurements that validate the theoretical results. It is shown that the raw dataset has inconsistencies, which reinforce the necessity to verify the truthfulness of the transactions. Encrypted measurements are necessary and sufficient to determine whether the computations and the measurements are correct.

Besides smart grids, several application areas can use the results of this book. ADC-Nets can be used to create several protocols provided with security, privacy, verifiability, scalability, reliability, efficiency, etc.

More important than efficiency, PPPs should enforce the security of customers' privacy by means of cryptography. Considering smart grids, PPPs are paramount for suppliers, for customers, and for the proper development of society.

Highlights

The following results of this book can be highlighted:

- This book presents limitations for all PPPs with data aggregations.
- This work introduces the concept of ADC-Nets.
- ADC-Nets have interesting properties including verification.
- This book presents an efficient ADC-Net.
- ADC-Nets are generalizations of additive homomorphic encryption.
- This work can be transferred to several other applications.

Acknowledgments

Many people have influenced my life for me to publish this book.

I'm thankful for all of them. Thank you very much!

Some people participated more intensively in this journey. Special thanks to my family and my Doktorvater.

I thank Prof. Max Mühlhäuser and Prof. Fengjun Li for the inspiration, talks, and corrections. I thank Renata and Lawrence Hamtil for proofreading.

I thank all my coauthors, especially those whose papers are references of this work. Thank you: Prof. Max Mühlhäuser, Prof. Leonardo Martucci, Prof. Renato Portugal, Prof. Johannes Buchmann, Prof. Franklin Marquezino, Prof. Nadia Nedjah, Prof. Pedro Lara, Dr. Raqueline Santos, Dr. Florian Volk, Dr. Filipe Beato, Dr. Denise Demirel, Dr. Albrecht Petzoldt, and Leon Böck. I also thank our editors and anonymous reviewers for their helpful comments.

I thank Prof. Neal Koblitz for interesting discussions about algebra, number theory, and cryptography.

Many friends of mine have made my life better. I thank all of them. In particular, three of them have been closer during this work. I thank Leonardo Martucci, Gino Brunetti, and Gerald Heinig.

Petrópolis, Brazil   Fábio Borges de Oliveira
February 2015

Contents

Part II   Contributions

List of Figures

List of Tables

List of Algorithms

List of Acronyms

ADC-Net Asymmetric DC-Net. iv, v, xiii, 7–10, 20, 30, 61, 80–84, 87, 90, 91, 101–103, 107, 108, 118, 127–129
AHEP additive homomorphic encryption primitive. xi, xiii, 7–10, 20–22, 26, 28–30, 65, 66, 83, 84, 101–108, 127, 128
AMI advanced metering infrastructure. 13, 14, 39
AMR automatic meter reading. 13

BSI German Federal Office for Security in Information Technology, the free translation of Bundesamt für Sicherheit in der Informationstechnik. 16

CSPRNG cryptographically secure pseudorandom number generator. 21, 83, 119

DLP Discrete Logarithm Problem. 67, 76, 78, 79, 81, 90–92, 97, 102, 103, 106, 132
DOE Department of Energy. 16

ECC Elliptic Curve Cryptography. 61, 66, 67, 76, 78
ECDLP Elliptic Curve Discrete Logarithm Problem. 8, 69, 77–79, 102–104, 122
EU European Union. xi, 13–15

GMT Greenwich Mean Time. 111, 112
GNFS general number field sieve. 78, 79, 132

HS homomorphic signature. 101–107

List of Abbreviations

List of Symbols

aggregated measurement (a_j) aggregated measurement of the phasor measurement unit (PMU) in the round j or expected consolidated consumption. xv, 39, 41, 44, 45, 63, 66, 72, 73, 75, 77, 87, 88, 90, 104, 113, 120
assignment $(\leftarrow)$ $a \leftarrow a+1$ means the assignment of $a+1$ to a. xxi, 29, 32, 34, 62, 65, 70, 71, 73, 75, 76, 85, 86, 88–90, 131, 132

bill $(b_i^\$)$ the amount charged in the invoice of the meter i, s.t. $b_i^\$ \overset{\text{def.}}{=} \sum_{j=1}^{\tilde{j}}$ Value $(m_{i,j})$ .. xxiii, 18, 21, 22, 26, 28, 32–34, 40, 41, 44, 45, 61, 72, 73, 75–77, 86, 89, 90, 97, 101, 104, 105, 108, 113, 127, 128
billed consumption (b_i) consumption in Watts registered in the invoice of the meter i, s.t. $b_{ii} \overset{\text{def.}}{=} \sum_{j=1}^{\tilde{j}} m_{i,j}$.. xi, xiii, 18, 39–41, 45, 46, 49–53, 55–58, 128
buying price $(\underline{p}_j)$ the buying price of a commodity in the round j. 62, 64

commitment function (Commit) a commitment scheme defined according to the protocol. 32, 34, 39, 71, 72, 77, 102
committed consolidated measurement $(\mathfrak{T}_i)$ consolidated committed measurement. 33, 34
committed measurement $(\mathfrak{M}_{i,j})$ committed measurement of $m_{i,j}$. xxiii, 32–34
complex numbers $(\mathbb{C})$ the set of the complex numbers. xxii
consolidated consumption (c_j) decrypted consolidated consumption of the measurements in the round j. xi–xiii, xxi, 10, 18, 21, 27–33, 39–46, 49–53, 55–59, 61, 73, 77, 82, 87, 104, 105, 108, 113, 117, 119, 120, 127, 128
consolidated monetary value $(c_j^\$)$ monetary value of the consolidated consumption of the measurements in the round j. 61, 63–65, 72–74, 77, 79, 85–88, 91–97, 101, 103–105, 107

decryption function (Dec) a decryption function defined according to the protocol. 19–22, 29, 31, 64, 65, 81–83, 86, 88

definition ($\overset{\text{def.}}{=}$) equal by definition. xxi, 18, 31–33, 56, 57, 63, 64, 71, 72, 74, 75, 81–83, 85–94

digital signature ($\mathfrak{S}_{i,j}$) digital signature of the meter i in the round j. xxiii, 32, 34, 44, 45, 71–73, 85–88, 102, 103

encrypted consolidated consumption ($\mathfrak{C}_j$) encrypted consolidated consumption of the measurements in the round j. 21, 27–30, 63–66, 73, 81, 82, 86–88, 90, 96, 119

encrypted measurement ($\mathfrak{M}_{i,j}$) encrypted measurement of the meter i in the round j, also known as ciphertext or cryptogram. iv, v, 6, 13, 18–22, 28, 29, 31–33, 39, 40, 45, 63–65, 78, 81, 83–90, 95, 105, 121, 128

encryption function (Enc) an encryption function defined according to the protocol. 18–22, 28, 29, 31, 63, 65, 81–83, 85

field ($\mathbb{F}$) a set $\mathbb{F}$ characterized by two abelian groups $(\mathbb{F}, \oplus)$ and $(\mathbb{F}, \odot)$ with $\odot$ distributing over $\oplus$. xii, 67–69, 71, 76, 77, 122

group ($\mathbb{Z}_p^*$) a multiplicative group of integers $\mathbb{Z}$ modulo p, where p is prime. 32

group ($(G, \circledast)$) a set with closure, associativity, identity, and inverse. 20, 67, 76, 77, 102, 105, 106

hash function (H) a secure hash function s.t. it behaves as a one-way function and has collision resistance. xii, 8, 31, 32, 63–66, 71, 72, 75–77, 81–83, 85, 90, 91, 97, 102, 103, 106, 107, 118–121, 123, 124

imaginary unit ($\imath$) the imaginary unit $\imath^2 = -1$ of complex numbers $\mathbb{C}$. 94, 95

integers ($\mathbb{Z}$) the set of the integer numbers. xii, xxii, 32, 69, 70, 76, 77, 81, 82, 85, 86, 90, 105, 106, 118

least common multiple function (lcm) function that returns the least common multiple. 29, 83

let ($\overset{\text{let}}{=}$) Let the state of the function be specified as follows. 31–33, 62–64, 71, 72

measurement ($m_{i,j}$) consumption measured by the meter i in the round j. iv, v, xi–xiii, xxi–xxiii, 3–11, 13, 16–21, 26–34, 39, 40, 43, 45, 46, 49, 50, 53–57, 61–66, 71, 72, 74–77, 81–91, 93, 94, 97, 103–108, 111–117, 128, 135

meter (i) meter, user, or customer identification. xi, xii, xxi–xxiii, 13–22, 27–29, 31–34, 39–42, 44–46, 50, 56, 57, 62–66, 71–78, 81–91, 93–95, 97, 102–107, 111–115, 117, 120, 123, 125, 128

monetary value (Value) function that returns the monetary value of a measurement $m_{i,j}$ of the meter i in the round j. xxi, 18, 41, 45, 62–65, 71, 72, 74, 76, 85, 87–89, 92, 94

natural numbers ($\mathbb{N}$) the set of the natural numbers. 54, 69
number of rounds ($\bar{j}$) total number of rounds j per bill $b_i^\$$. xi, xxi, 18, 32–34, 45, 46, 49–57, 75, 76, 86, 89, 90, 106, 111, 128
number of users ($\bar{i}$) total number of meters, users, or customer identifications. xi, 18, 21, 27–29, 31–33, 45, 46, 49–57, 64–66, 71–73, 77, 81–88, 90, 91, 93–97, 103, 106–108, 111, 120, 127, 128

open function (Open) a function defined according to the protocol to return true **iff** the values are correct with respect to committed measurements $\mathfrak{N}_{i,j}$. 33, 34, 72, 73, 75–77, 87

privacy component (PC_i) a plug-in privacy component PC_i attached to the meter i. xxiii, 32–34

real numbers ($\mathbb{R}$) the set of the real numbers. 54, 68
round (j) round identification. xii, xiii, xxi–xxiii, 18–21, 27–29, 31–34, 40, 41, 44–46, 49, 50, 52, 56, 57, 62–65, 71–77, 81–83, 85–91, 93, 94, 97, 103, 104, 106, 107, 111–114, 117–120, 123, 124

selling price ($\bar{p}_j$) the selling price of a commodity in the round j. 62, 64

sign function ($\text{sgn}(m_{i,j})$) $\text{sgn}(m_{i,j}) = \begin{cases} -1 & \text{if } m_{i,j} < 0 \\ 0 & \text{if } m_{i,j} = 0 \\ 1 & \text{if } m_{i,j} > 0 \end{cases}$. 62, 64

signature function (Sign) a secure function that returns a digital signature $\mathfrak{S}_{i,j}$. 34, 71, 85

verify function (Verify) the function that verifies the signature. 34, 73, 86, 88

whether ($\stackrel{?}{=}$) the values are correct whether the equation holds. 41, 72, 74–76, 88, 89

Part I
Foundations

For centuries, the human being had burned fossil fuel and only stopped before the end of the resources because the burning caused global warming and clime change with extreme seasons. Not the four seasons that we know today, but for instance, periods of intense rain with floods in the winter or snowstorms in the summer. To survive, they burned increasingly more, resulting in a more hostile environment. Trying to stop the snowball effect, they constructed nuclear power plants and used nuclear fusion. Unfortunately, earthquakes and tsunamis caused tragic events. In addition, the technology demanded always more electricity and each human being needed a huge amount of electricity to survive in such extreme environment. The costs of living in the earth were becoming more expensive than in Mars.

An army of scientists had worked with The Cloud on the sustainability problem. Thousands of supercomputers formed The Cloud, which helped them to use the electricity with extreme efficiency, but The Cloud itself required too much energy for the available resources. Many cities started shutting down their supercomputers when The Cloud understood how to find the solution. On July 9, 2184, what was predicted in theory and simulations became reality. Solar energy started being captured, converted, stored, and utilized on a global scale. Billions of highly connected solar panels covered the earth in approximately a tenth of the distance to the Moon. Invisible photovoltaic cells yielded all electricity demanded.

"Awesome! The science seems a sudden miracle and the technology seems a common magic!" Said an astonished post-doc. "We have all the necessary energy to control the weather. Indeed, we have much more energy than we need to run many planets. Even better, all energy is renewable and for free! Forever and ever!"

It was a small talk in a graduation reception. However, their professor was in the circle and raised a subtle and ironic smile.

"Yes, professor. Forever is a long time. I mean billions of years until the end of the solar system. Now, you agree on it. Don't you?"

—Inspired by Isaac Asimov, *The Last Question*

Chapter 1
Introduction

Abstract Information and communication technologies (ICTs) have become ubiquitous. Particularly, the Internet has become the global supply network for information and communication in the form of byte streams. Increasingly, ICT is used to control other supply networks, e.g., for electricity, gas, and heating. The aim is to make networks "smarter," in particular to provide more efficiency and flexibility in the demand-supply loop, which would greatly benefit from "real-time" consumption measurements sent from consumers to suppliers, and from "intelligent" appliances that would adjust consumption to the tidal waves of energy abundance or scarcity. However, the introduction of ICTs with fine-grained measurements in particular brings very serious privacy risks. In present supply networks, risks can be mitigated only by means of very restrictive legal measures. In this light, the present book aims at providing a high level of privacy while enabling frequent measurements by means of a Privacy-Preserving Protocol (PPP). A PPP along with its algorithms keeps users' privacy secure.

Keywords Measurements • Non-smart grid • Disaggregation • Consumption • Privacy • Research topic • Renewable energy • Benefits

This book addresses a threat represented by smart metering systems, which are able to collect frequent measurements of consumption data. Specifically, smart meters are the most famous components of smart metering systems. Suppliers have installed smart meters in customers' properties to measure their consumption. Figure 1.1 depicts a gathering of measurements in a non-smart grid, i.e., non-smart meters measure the amount of electricity consumed by customers and an employee goes house by house collecting non-frequent measurements.

ICTs enable smart meter to collect measurements in a much more efficient and reliable way. However, many customers realized that an attacker could measure remotely at any time. Consequently, the attacker could infer private information. Indeed, smart meters can reveal detailed information about customers' private life. Fine-grained consumption data collection can even disclose what customers

---

This book is an extended version of a Ph.D. thesis [Fábio Borges de Oliveira: On Privacy-Preserving Protocols for Smart Metering Systems, Technische Universität, Darmstadt, 2015.]

© Springer International Publishing Switzerland 2017   3
F. Borges de Oliveira, *On Privacy-Preserving Protocols for Smart Metering Systems*,
DOI 10.1007/978-3-319-40718-0_1

Fig. 1.1 A supplier's employee collecting measurements measured by non-smart meters

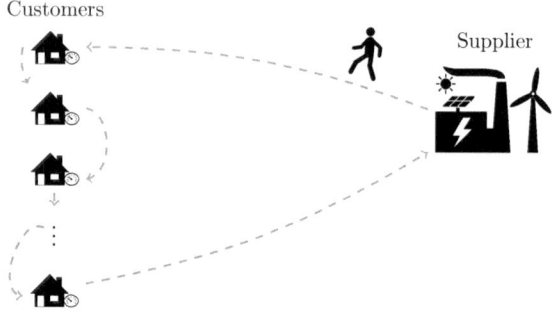

are watching on their TVs [14]. The amount of information disclosed depends on the level of granularity [19]. If the measurements are collected on a monthly basis, attackers can use previous measurements to deduce when a customer usually vacations. Therefore, they can infer which customers are on vacation this month. A measurement might have the consumption of several appliances in an interval. The disaggregation of such measurement can be done with a technique called non-intrusive load monitoring (NILM) [23], which can be used to infer private information of inhabitants. Nevertheless, NILM has a wide range of applications, including the shipping industry [11]. The technique is still in development, but one can already find an open source toolkit for NILM [2]. Figure 1.2 depicts the disaggregated consumption of a customer on Sunday, May 1, 2011. The data used in Fig. 1.2 belongs to the Reference Energy Disaggregation Data Set known as REDD [17]. Note that the total consumption given by a measurement is the sum of the curves in Fig. 1.2. In addition, the curve given by the sum is very similar to Fig. 1.2.

It is difficult to mask signatures carried in the fine-grained consumption data as well as it is difficult to hide private information even for coarse-grained consumption data. One can use a battery to protect privacy. However, privacy protection is directly proportional to the volume and price of the battery. Currently, only expensive large-volume batteries can give a very good level of privacy protection. Therefore, using a battery is still not a solution. Smart meters measure the consumption of a commodity delivered by flow. Examples of these are electricity, water, natural gas, and heating. Thus, a supplier is the intended recipient of measurements. Besides electricity, NILM can be used for other commodities [13]. Among them, the most interesting privacy issue lies on electricity consumption, because the storage of large amount of energy is still expensive. Thus, customers should consume the electricity in the same instant of its generation. In contrast, a privacy-aware customer of a water supplier can store water to cancel the smart meter intrusion. Similarly, a customer can store natural gas and heating. Since electricity cannot be stored in large quantities, most of the literature on PPPs addresses problems about power grid. Therefore, this book focuses more on energy suppliers. In general, a smart grid is seen as the enhancement of a power grid with new technologies. In other words, it is seen as the evolution [4] of a power grid. Despite its association with the power grid [7], a

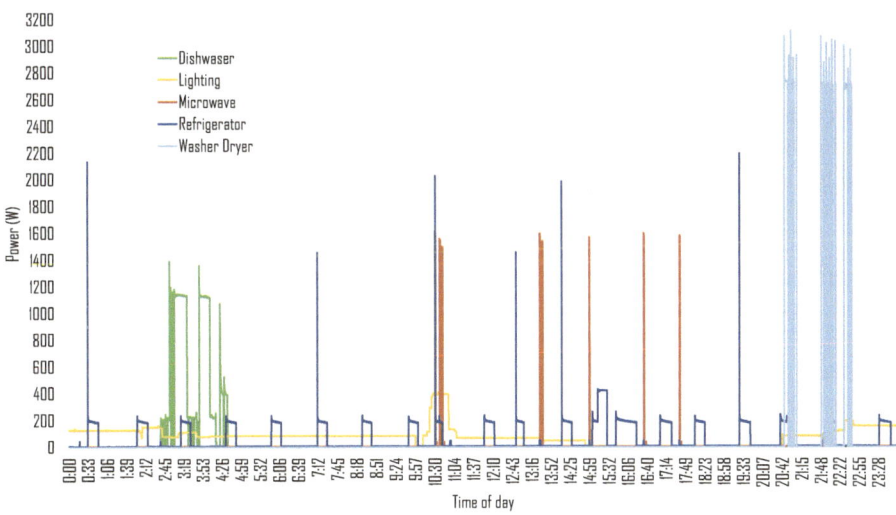

Fig. 1.2 Disaggregated electricity consumption on Sunday, May 1, 2011

smart grid is better defined as a network of people, computers, and sensors in public infrastructures that monitors and manages the usage of commodities. In contrast to a first generation computerized supplier, a smart grid can be constructed to control the flow of a commodity. Consequently, electric vehicles, smart homes, and even the Internet can be seen as part of the smart grid.

Although the literature provides many proposals for PPP, few of them meet the requirements identified in this work.

1.1 Motivation

With the ubiquity of ICTs, companies such as Google and Microsoft were investing to collect customers' measurements. However, many customers were concerned about privacy issues [1]. In fact, information is power and money, and massive leaks of privacy might eventually lead to a surveillance society and compromise democracy [15]. Despite the huge amount of data and information provided by smart meters, Google PowerMeter was discontinued in 2011 and Microsoft Hohm in 2012. Researchers were wondering whether smart meters would be a research topic for future years. In addition, the requirements for PPPs were unclear. Smart meters would not be an essential part of a smart grid scenario. Certainly, remote measurements reduce costs with employees, but such savings would not be worth due to the increasing concerns regarding security and privacy in smart grids.

The initial motivation for energy suppliers to deploy smart meters was to charge customers with dynamic pricing [12]. Thus, suppliers would match consumption with generation to achieve power load balancing by means of demand response, which is defined as the changes in their customers' consumption patterns that happen in response to changes in the price. Consequently, society can save construction of expensive power plants that run only during consumption peaks. In some countries, such peaking power plants need fuels to burn. Therefore, avoiding them implies a reduction of CO_2. Certainly, renewable energy sources are the key to the reduction of CO_2 emissions. Despite the fact that solar photovoltaic and wind are forms of renewable energy, they are unstable. To use them, electric vehicles appear as an opportunity to store the energy in generation peaks and to supply the power network according to customers' needs [22]. For example, in the United States of America (USA), energy suppliers are ready to handle millions of electric vehicles without changing their generation and transmission infrastructure [21]. Similarly, the electrified German railway network could be equipped with batteries to store the excess energy and to act as supplier during scarcity.

Currently, some energy suppliers need to smooth peaks due to consumption and generation. Smart meters are an option to smooth the peaks. Imbalances between supply and demand cause high prices. In addition, too much energy generation leads to negative prices [20]. Therefore, society could profit from smart meters.

On the one hand, smart meters can yield many benefits for society, including environmental and economic. On the other hand, they cause too much privacy intrusion [1]. The solution for this dilemma can be achieved with a PPP. In this case, algorithms run on smart meters and on supplier computers to protect customers' privacy when smart meters send their encrypted measurements to their supplier.

Currently, many suppliers are deploying smart meters, and many countries are requiring their deployment. Smart meters are here to stay, and researchers are discovering new applications that require them, for instance, for overload detection in old transformers [18]. They will be in the market because there are strong economic and ecological reasons. Furthermore, this book presents new reasons for smart meters. Due to privacy issues, they should be deployed with a PPP, which can safeguard the proper development of society.

1.2   Research Questions

This research was initiated to evaluate privacy threats and requirements in smart grid scenarios for designing PPPs to protect customer privacy. The privacy threats were quickly identified in the literature. Therefore, the following research questions were raised:

1. What are the reasons to deploy smart meters?
2. What are the requirements for PPP?
3. How often should the measurements be collected?
4. How many smart meters should compose the aggregation?
5. What are the properties of the protocols?

The privacy issues depend on how often the smart meters send their measurements to their supplier, i.e., the privacy issues are close to minimal with infrequent measurements. In some countries, suppliers normally collect measurements on a monthly basis. In some others, suppliers normally collect them on a yearly basis. This is the case of Germany, which has a law adapted for smart meters but does not allow suppliers to create profile of customers with consumption upto 100,000 kw h.[1] Germans have the concept of annual consumption—*Jahresverbrauch* in German— which enables customers to pay monthly a fixed estimated amount. If the smart meters send the measurements yearly, the supplier receives the same information that a non-smart grid provides and attackers have no idea when customers have vacation. The problem appears when someone can request measurements anytime, or when the smart meters send them on a shorter fixed interval.

The requirements for PPPs depend on the frequency of the measurements. With very few measurements, customers might wish to protect their measurements only from neighbors. Perhaps, the suppliers can receive the measurements on a weekly basis without a big problem. In contrast, frequent measurements represent a threat for customers' privacy.

Similarly, the number of smart meters influences the selection of the best PPP. Efficiency in terms of processing time and communication depends on this number, which influences the scalability of some protocols [8].

If more than one PPP meets the requirements, the most efficient is better. Efficiency is one determinant factor for the price of smart meters. Moreover, small savings in processing time may imply billions of smart meters that have smaller energy consumption per measurement with cheaper hardware. The more efficient the PPP, the lower the price for privacy.

Besides these main questions, many others have been raised. For example, could we use quantum cryptography or post-quantum cryptography to construct a PPP?

1.3 Contributions

This book has two major contributions. One is the presentation of state-of-the-art PPPs in a smart grid scenario. The other is the theoretical contribution for research on privacy. Specifically, introducing the concept of Asymmetric DC-Nets (ADC-Nets) and showing the relations between ADC-Nets, Symmetric DC-Nets (SDC-Nets), and additive homomorphic encryption primitives (AHEPs).

The first two research questions are addressed in Chap. 4, which presents scenarios where smart meters should send their measurements as frequently as possible. It also presents the minimum requirements for PPPs. The third and fourth research questions are addressed in Chap. 5, which presents limitations for all PPP based on data aggregation. It also presents that the number of smart meters in

---

[1] In German: https://www.gesetze-im-internet.de/stromnzv/BJNR224300005.html.

the aggregation should be equal to the sum of the measurements to maximize the number of possibilities. However, Chap. 5 does not present a minimum number, which depends on several variables. The fifth research question is addressed in Chap. 6. As a result, PPPs should have the properties of ADC-Nets.

This work has proposed several PPPs, e.g., [3, 6, 7]. Some proposals have more than one PPP [5, 10]. This book presents four interesting PPPs.

1.3.1   High Level Explanation of the Selected PPPs

Each PPP has a different cryptographic primitive and a discussion about security, privacy, and performance.

1.3.1.1   PPP1

The first selected PPP uses in-network aggregation with SDC-Net to behave as AHEP. Hence, each smart meter needs to compute only one "one-way function" to encrypt the measurements. Therefore, this raises the conjecture that PPP1 is as fast as possible in the smart meter side. PPP1 uses a hash function as a "one-way function," but techniques of symmetric cryptography might be used to speed it up [9]. PPP1 only provides aggregation and is an improved part of a protocol called iKUP [5].

1.3.1.2   PPP2

PPP2 uses commitment over elliptic curves, which are very famous for being a fast cryptographic primitive, because the best algorithms to solve the Elliptic Curve Discrete Logarithm Problem (ECDLP) need much more time than the best algorithms to solve the Integer Factorization Problem (IFP). Depending on the number of measurements and their size, suppliers can use brute force to "decrypt" the aggregation. PPP2 uses one scalar multiplication to "encrypt" and another to encode the measurement. If we assume that it cannot be done with one scalar multiplication, PPP2 is the fastest based on the ECDLP. PPP2 has many interesting features. For example, billing information can be verified, suppliers can identify smart meters with communication problems, it is free of trusted third party (TTP), etc. However, PPP2 also has a drawback. It is not scalable on the supplier side. PPP2 is also an improved part of iKUP [5].

1.3.1.3   PPP3

PPP3 uses an ADC-Net, which is also introduced in the work of this book [8]. PPP3 has all the features that PPP2 has. In addition, PPP3 is scalable and even faster than PPP2. Indeed, only PPP1 is faster than PPP3 on the smart meter side. On the supplier side, PPP3 has constant time with respect to the number of smart meters, while PPP2 increases the processing time with the number of smart meters. PPP3 is the most suitable PPP.

1.3.1.4   PPP4

PPP4 is a preliminary PPP based on quantum cryptography. The measurements are hidden in a quantum entanglement with a shift phase operator. Besides the use of quantum mechanics, PPP4 does not need a quantum computer. It can use commercially available products, but it is still expensive. Therefore, its deployment depends on technological developments.

1.3.2   Summary of the Results

Besides the state of the art for privacy in smart grids, this book presents several interesting PPPs. Four of them are improvements of PPPs developed in this work and already published. The theoretical analysis agrees with the simulation, which is done to cover the majority of the PPPs. The unique selected PPP that does not have simulation is based on quantum mechanics. One of them uses the concept of ADC-Nets, which advances the concept of SDC-Nets. ADC-Nets have many new interesting properties, cf. Sect. 6.4.1.

Besides the results already published, this book presents new results that were only presented at seminars. The main unpublished results are the reasons for smart meters to collect frequent measurements, privacy quantification of the aggregation size, and the result that each AHEP is a particular case of an ADC-Net. The last result is a method to transform AHEP in an ADC-Net.

1.4   Outline

The chapters can be read in any order. The acronyms and symbols are reintroduced in each new chapter, and at the end of this book, there is a glossary, while at the beginning, there is a list of acronyms, a list of abbreviations, and a list of symbols. Thus, one expert can select a PPP in Chap. 3 or Chap. 6 and can read it without paying attention to the previous protocols or chapters. However, for anyone to whom this research is new, this book is organized in an order that simplifies understanding. The remainder of this book is structured as follows:

Chap. 2 contextualizes the researches and projects on smart grids around the world. In addition, it presents the terms used to describe PPPs, a security model, and a privacy model used in many PPPs for smart grid scenarios.

Chap. 3 surveys the most relevant privacy-enhancing technologies for this work. They are split into solutions with a strong disadvantage, and solutions with strong advantages, which are inspirations for the four PPPs introduced in this work.

Chap. 4 clarifies the problem showing the new reasons for smart meters and data aggregation. Moreover, it also shows a scenario that requires remote measurements to be performed as frequently as possible. Since there are economic reasons for the requirements, it is believed that smart meters will be ubiquitous eventually. This is necessary because the initial motivation for smart meters was to charge customers with dynamic pricing [12], but smart grids can achieve this without frequent remote measurements [16]. Since one protocol in the related work shows that smart grids can have dynamic pricing without smart meters sending their measurements to their supplier, this book assesses the importance of aggregated data and presents scenarios that require such data. At the end of the chapter, a section presents the minimum requirements for PPPs. Specifically, Section 4.2 presents four requirements that each PPP for smart metering systems should fulfill. However, the majority of the protocols found in the literature have addressed only one requirement, namely consolidated consumption by aggregation of measurements [3, 5].

Chap. 5 quantifies the correspondence between aggregation size and risk of privacy leakage as a function of several variables. This chapter divides the search for the individual measurements into algebraic and probabilistic properties derived from all PPPs that provide aggregation. The former results in an error-correcting code for suppliers and leads to equations that reveal the number of possibilities for an attacker. Maximum security is achieved when the variables have the same value. The latter shows that an attacker can always find the most probable individual measurements in few steps. The difficulty for the attacker is to define what is more probable. This chapter analysis is independent of protocol and is valid for all PPPs that satisfy the minimum requirements.

Chap. 6 starts with the definition of a function that converts the consumption into a monetary value. The four PPPs use this function. This technique simplifies the protocols in comparison with their previous versions already published in scientific papers. The chapter presents the innovative protocols from the simplest to the most complex based on quantum cryptography. PPP1 has constant time on the meter side. Excluding the fastest PPP1, PPP2 and PPP3 are increasingly faster than others are when the level of security increases. PPP4 has the problem of keeping and accessing the quantum information that are common to other algorithms based on quantum mechanics. Currently, PPP3 is the most recommended for smart grids. It uses an ADC-Net, which is introduced in Section 6.4.1. In addition, the relation of the ADC-Net protocol with AHEPs is also presented. As a protocol result, customers can keep their private life secure without having to trust any institution.

Chap. 7 compares different strategies used for the four selected PPPs—among others—regarding features in the minimum requirements, verification property, security, privacy, and performance, which is split into processing time and communication costs. The comparison is theoretical and based on complexity analysis. Its last section compares SDC-Nets, AHEPs, and ADC-Nets, showing the benefits of ADC-Nets in comparison with other techniques. The chapter clarifies why ADC-Nets are more suitable to PPPs and why PPP3 is more suitable to smart grids than the other selected protocols. Briefly, PPP1 can be much faster on the smart meter side, but it does not have the features that PPP3 has.

Chap. 8 validates the theoretical analysis of processing time with simulation of state-of-the-art PPPs in a parallel computer using as input more than one hundred million real-world measurements, which were collected with a fixed interval of 30 min by more than six thousand smart meters during one and a half years. The analysis of the raw dataset detects anomalies, which reinforce the idea that PPPs should enable verification. The chapter also presents the tools used in the simulation and the parameters used in the implemented algorithms.

Chap. 9 concludes this book in four steps, namely, recapitulating it, highlighting the main results, presenting new perspectives, and finalizing the conclusion. The first connects the whole book. The second emphasizes the results. The third prospects new research topics from the presented results. The fourth synthesizes the importance of this book.

References

1. L. AlAbdulkarim, Z. Lukszo, Impact of privacy concerns on consumers' acceptance of smart metering in the Netherlands, in *2011 IEEE International Conference on Networking, Sensing and Control (ICNSC)* (2011), pp. 287–292. doi:10.1109/ICNSC.2011.5874919
2. N. Batra et al., NILMTK: an open source toolkit for nonintrusive load monitoring, in *Proceedings of the 5th International Conference on Future Energy Systems*. e-Energy '14 (ACM, Cambridge, 2014), pp. 265–276. isbn:978-1-4503-2819-7. doi:10.1145/2602044.2602051. http://doi.acm.org/10.1145/2602044.2602051
3. F. Borges, L.A. Martucci, iKUP keeps users' privacy in the smart grid, in *2014 IEEE Conference on Communications and Network Security (CNS)* (2014), pp. 310–318. doi:10.1109/CNS.2014.6997499
4. F. Borges, M. Mühlhäuser, EPPP4SMS: efficient privacy-preserving protocol for smart metering systems and its simulation using real-world data. IEEE Trans. Smart Grid **5**(6), 2701–2708 (2014). doi:10.1109/TSG.2014.2336265. http://dx.doi.org/10.1109/TSG.2014.2336265
5. F. Borges, L.A. Martucci, M. Mühlhäuser, Analysis of privacy-enhancing protocols based on anonymity networks, in *2012 IEEE Third International Conference on Smart Grid Communications (SmartGridComm)* (2012), pp. 378–383. doi:10.1109/SmartGridComm.2012.6486013
6. F. Borges, A. Petzoldt, R. Portugal, Small private keys for systems of multivariate quadratic equations using symmetric cryptography, in *XXXIV CNMAC - Congresso Nacional de Matemática Aplicada e Computacional*. Águas de Lindóia - SP (2012), pp. 1085–1091. http://www.sbmac.org.br/eventos/cnmac/xxxiv_cnmac/pdf/578.pdf

7. F. Borges et al., A privacy-enhancing protocol that provides innetwork data aggregation and verifiable smart meter billing, in *2014 IEEE Symposium on Computers and Communication (ISCC)* (2014), pp. 1–6. doi:10.1109/ISCC.2014.6912612

8. F. Borges, J. Buchmann, M. Mühlhäuser, Introducing asymmetric DC-Nets, in *2014 IEEE Conference on Communications and Network Security (CNS)* (2014), pp. 508–509. doi:10.1109/CNS.2014.6997528

9. F. Borges, R.A.M. Santos, F.L. Marquezino, Preserving privacy in a smart grid scenario using quantum mechanics. Secur. Commun. Networks, n/a (2014). issn:1939-0122. doi:10.1002/sec.1152. http://dx.doi.org/10.1002/sec.1152

10. F. Borges, F. Volk, M. Mühlhäuser, Efficient, verifiable, secure, and privacy-friendly computations for the smart grid, in *2015 IEEE Power Energy Society Innovative Smart Grid Technologies Conference (ISGT)* (2015), pp. 1–5. doi:10.1109/ISGT.2015.7131862

11. T. DeNucci et al., Diagnostic indicators for shipboard systems using non-intrusive load monitoring, in *2005 IEEE Electric Ship Technologies Symposium* (2005), pp. 413–420. doi:10.1109/ESTS.2005.1524708

12. P. Fox-Penner, *Smart Power: Climate Change, the Smart Grid, and the Future of Electric Utilities* (Island Press, Washington, DC, 2010). isbn:9781597268097

13. J. Froehlich et al., Disaggregated end-use energy sensing for the smart grid. IEEE Pervasive Comput. **10**(1), 28–39 (2011). issn:1536-1268. doi:10.1109/MPRV.2010.74

14. U. Greveler, B. Justus, D. Löhr, Multimedia content identification through smart meter power usage profiles, in *Computers, Privacy and Data Protection (CPDP 2012)* (2012)

15. J. Holvast, History of privacy. English, in *The Future of Identity in the Information Society*, ed. by V. Matyáš et al., vol. 298. IFIP Advances in Information and Communication Technology (Springer, Berlin, Heidelberg, 2009), pp. 13–42. isbn:978-3-642-03314-8. doi:10.1007/978-3-642-03315-5_2. http://dx.doi.org/10.1007/978-3-642-03315-5_2

16. M. Jawurek, M. Johns, F. Kerschbaum, Plug-In Privacy for Smart Metering Billing, in *Privacy Enhancing Technologies: Proceedings of 11th International Symposium, PETS 2011, Waterloo, ON, July 27–29, 2011*, ed. by S. Fischer-Hübner, N. Hopper (Springer, Berlin, Heidelberg, 2011), pp. 192–210. isbn:978-3-642-22263-4. doi:10.1007/978-3-642-22263-4_11. http://dx.doi.org/10.1007/978-3-642-22263-4_11

17. J.Z. Kolter, M.J. Johnson, REDD: a public data set for energy disaggregation research, in *Workshop on Data Mining Applications in Sustainability (SIGKDD), San Diego, CA*, vol. 25 (2011), pp. 59–62

18. K.D. McBee, M.G. Simoes, General smart meter guidelines to accurately assess the aging of distribution transformers. IEEE Trans. Smart Grid **5**(6), 2967–2979 (2014). issn:1949-3053. doi:10.1109/TSG.2014.2320285

19. A. Molina-Markham et al., Private memoirs of a smart meter, in *Proceedings of the 2nd ACM Workshop on Embedded Sensing Systems for Energy-Efficiency in Building*. BuildSys '10 (ACM, Zurich, 2010), pp. 61–66. isbn:978-1-4503-0458-0. doi:10.1145/1878431.1878446. http://doi.acm.org/10.1145/1878431.1878446

20. M. Nicolosi, Energy efficiency policies and strategies with regular papers. Energy Policy **38**(11), 7257–7268 (2010). issn:0301-4215. doi:10.1016/j.enpol.2010.08.002. http://www.sciencedirect.com/science/article/pii/S0301421510005860

21. D. Novosel, V. Rabl, J. Nelson, A report to the U.S. DOE: IEEE shares its insights on priority issues [leader's corner]. IEEE Power Energ. Mag. **13**(2), 6–12 (2015). issn:1540-7977. doi:10.1109/MPE.2014.2374971

22. C. Rottondi, S. Fontana, G. Verticale, A privacy-friendly framework for vehicle-to-grid interactions. English, in *Smart Grid Security*. ed. by J. Cuellar. Lecture Notes in Computer Science (Springer International Publishing, Cham, 2014), pp. 125–138. isbn:978-3-319-10328-0. doi:10.1007/978-3-319-10329-7_8. http://dx.doi.org/10.1007/978-3-319-10329-7_8

23. Z. Wang, G. Zheng, Residential appliances identification and monitoring by a non-intrusive method. IEEE Trans. Smart Grid **3**(1), 80–92 (2012). issn:1949-3053. doi:10.1109/TSG.2011.2163950

Chapter 2
Background and Models

Abstract This chapter contextualizes the role of smart meters in smart grid initiatives around the world to show that the smart grid concept goes beyond energy supplier modernization. In addition, this chapter presents the security model and the privacy model for Privacy-Preserving Protocols (PPPs). Security is ensured by means of cryptography, and privacy is protected by aggregation of encrypted measurements.

Keywords Initiatives • Concept • Security • Privacy • Aggregation • Maps • Cryptography • Aggregation

2.1 Smart Grids Around the World

On the Internet, one can find many projects and governmental sites about smart grids. Smart Metering Projects Map in Google Maps[1] provides a good visualization of the number and distribution of smart grid initiatives around the world. Figure 2.1 shows a screen-shot of the map.[2] In addition, Fig. 2.2 gives us a zoomed-in view of smart grid initiatives in the European Union (EU). A triangle indicates a trial or pilot, and a circle indicates a project. The colors red, green, and blue represent initiatives for electricity, gas, and water, respectively. Red is the dominant color, thus indicating that the majority of the initiatives are directed to electricity. The initiatives are also classified as automatic meter reading (AMR), advanced metering infrastructure (AMI), and smart grid. The first aims mainly to collect measurements and send them to suppliers. The second aims to transform the metering systems into microcomputers connected in networks. The third aims to use additional technologies. The AMI is the new terminology and goes beyond AMR. In terms of technology, the idea of AMR is old [8]. However, it was renewed with the AMI, which integrates new features like remote control and two-way communication [6].

---

[1]http://maps.google.com/maps/ms?ie=UTF8&oe=UTF8&msa=0&msid=1155193110583675343 48.0000011362ac6d7d21187.

[2]On January 1, 2015.

© Springer International Publishing Switzerland 2017   13
F. Borges de Oliveira, *On Privacy-Preserving Protocols for Smart Metering Systems*,
DOI 10.1007/978-3-319-40718-0_2

Fig. 2.1 Smart Metering Projects Map—Google Maps

Fig. 2.2 Smart Metering Projects Map in EU—Google Maps

A smart grid can have even more than AMI, for instance, phasor measurement units (PMUs), distributed generation, and smart inverters. Information about interesting features of smart inverters can be found in [5].

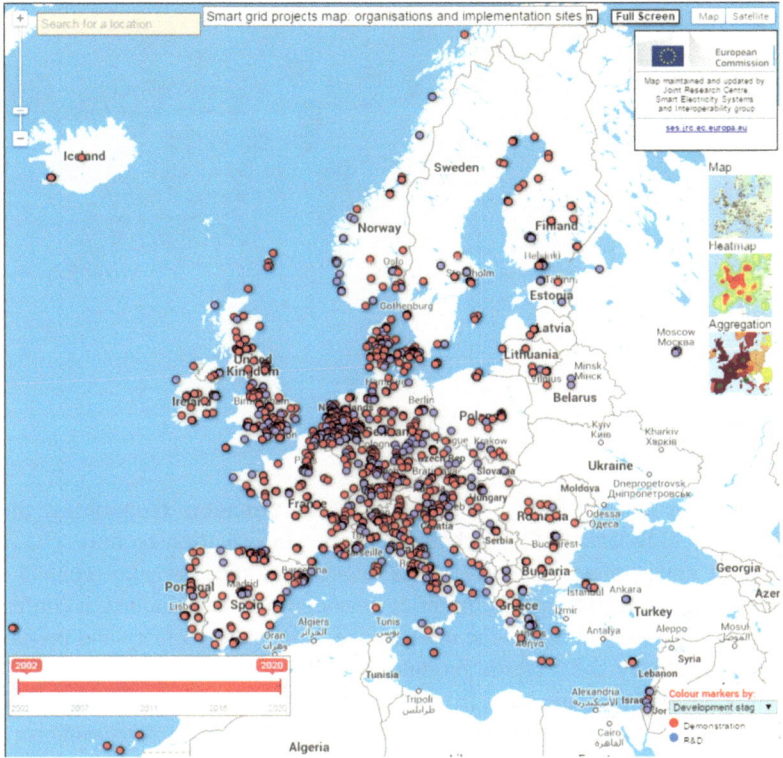

Fig. 2.3 Official Smart Metering Project Map in EU

Currently, many initiatives are taken to create smart grids. The EU aims to install smart meters in 80 % of households by 2020.[3] Figure 2.3 shows a screen-shot[4] of the official map[5] generated by European Commission's in-house science service. The map is interactive and can show information about initiatives associated with the EU outside of Europe, for instance, in the America. The EU also aims to reach at least 80 % reduction of greenhouse gas emission by 2050 in comparison with 1990 levels.[6] A survey of regulations in the EU electricity market may be found in [11]. This chapter does not present political regulation in depth because of its ephemeral nature. In particular, countries need to adapt their laws for modern smart metering systems. Particularly, Germany has made efforts to increase substantially the share of renewable and private energy production. Such efforts are known as *Energiewende*.

---

[3]Directive 2012/27/EU of 25 October 2012 published on the Official Journal L No.315, 25 Oct 2012.

[4]On January 1, 2015.

[5]http://ses.jrc.ec.europa.eu/.

[6]Energy roadmap 2050—EU—doi:10.2833/10759.

Fig. 2.4 Official Smart Metering Project Map in USA

The German Federal Office for Security in Information Technology, the free translation of Bundesamt für Sicherheit in der Informationstechnik (BSI), has defined that smart meter gateway has a secure module and may control many metering devices of different commodities in a neighborhood. Indeed, it controls the communication and centralizes the intelligence. To ensure security and privacy, the smart meter gateway has a secure module, like a Trusted Platform Module (TPM), and aggregates the measurements from many metering devices to ensure privacy, cf. Sect. 2.2.3. In addition, the prescription of the German BSI *Schutzprofil* for smart meters mitigates the risks by means of very restrictive legal measures.

The Department of Energy (DOE) of the United States of America (USA) also presents a map of investments in smart grids. Figure 2.4 shows a screen-shot of the official map[7] generated by the DOE.[8] In the USA, smart grid is a term applied to the power grid modernization due to its aging. Electrification is recognized as the greatest achievement of impact on quality of life and as uniquely critical system [7].

In contrast to the BSI model that can work with multiple commodities, the National Institute of Standards and Technology (NIST) has focused its standards on smart grid scenarios for energy suppliers. The NIST Framework and Roadmap for Smart Grid Interoperability Standards[9] presents a conceptual reference model to describe the interaction between the information network and the electric power network. In this standard, seven domains are defined as below.

Customers are the electricity consumers in the power network who may also be small generators for some periods.

Markets are parties involved in the electricity markets.

---

[7]On January 1, 2015.

[8]https://www.smartgrid.gov/recovery_act/project_information.

[9]NIST Special Publication 1108R2.

Service Providers are the organizations that provide services to customers and
    suppliers.
Operations is a domain in which actors manage the electric flow.
Bulk Generation is the set of large-scale electricity generators.
Transmission indicates the corporations responsible for the transmission of elec-
    tricity in high voltage from distant power plants to distribution networks.
Distribution indicates the corporations responsible for distributing the electricity
    between the customers in the distribution power network.

NIST is one of the pioneers in smart grid privacy issues. In 2010, the guideline
for Privacy and the Smart Grid[10] drew attention to the fact that the energy supplier
can identify when customers turned on and turned off their appliances. The USA
have made strong investment in smart meters and aim to have almost 52 million
customers equipped with smart meters by 2015 [4]. In 2012,[11] suppliers in USA
already had more than 43 million smart meters installed.

2.2 Security and Privacy Models

Security and privacy models for smart grid scenarios require the definition of some
terminology. A PPP should have a usual secure model, but its privacy model goes
further than the secure model. In fact, this section goes from the basis of the security
to lay down the bases for a privacy model.

2.2.1 Terminology in PPPs

This book uses some specific terms as listed below. Others may be found in the
Glossary at the end of this book or at the beginning in List of Acronyms, List of
Abbreviations, or List of Symbols.

User is an abstraction of a customer with a smart meter running a PPP with a
    supplier. The user may buy or sell a commodity.
Supplier is an abstraction of bulk generator, transmission, distribution, operations,
    markets, and service providers.
Meter is an abbreviation of smart meter, which lies in a customer's property. Its
    function is to collect measurements from a commodity flow and to report them
    through an information network to a supplier. Meters can communicate in many
    ways, e.g., using wireless, power line communication, or Internet Protocol (IP).

---

[10]NISTIR 7628 Guidelines for Smart Grid Cyber Security: Vol. 2, Privacy and the Smart Grid.
[11]http://www.eia.gov/tools/faqs/faq.cfm?id=108&t=3.

Round (or round of measurement) is a period in which a supplier receives the
  encrypted measurements from every meter i. Normally, the meters considered
  in one round belong to the same neighborhood. The measurements are collected
  in a fixed interval or by a request of the supplier.

Measurement is the measured consumption or generation in watts collected by a
  meter i in the round j, and it is denoted as $m_{i,j}$. Normally, the interval between
  rounds is assumed to be short.

Consolidated consumption is the sum of the measurements $m_{i,j}$ in the round j, and
  it is denoted as c_j. Thus, c_j is the total of energy consumption or generation
  reported by all meters during one round j to their supplier, i.e.,

$$c_j \overset{\text{def.}}{=} \sum_{i=1}^{\tilde{i}} m_{i,j},$$

where $\tilde{i}$ is the number of meters in the aggregation.

Bill is a monetary consumption value of an invoice with respect to the electricity
  consumption or generation in a period, and it is denoted by $b_i^{\$}$, i.e.,

$$b_i^{\$} \overset{\text{def.}}{=} \sum_{j=1}^{\tilde{j}} \text{Value}\left(m_{i,j}\right),$$

where $\tilde{j}$ is the number of rounds until the billing process and $\text{Value}\left(m_{i,j}\right)$ is
  a function that transforms the measurements from watts into a monetary value
  with a price that floats over the time. Thus, the electricity has a time-based
  pricing.

Billed consumption is the balance of the consumption and generation in watts
  registered in the invoice of the meter i. This balance is denoted as b_i and given by

$$b_i \overset{\text{def.}}{=} \sum_{j=1}^{\tilde{j}} m_{i,j}.$$

Note that the measurements can be positive or negative depending on whether
there is consumption or generation. In addition, the time-based pricing might be
different in buying or selling. Normally, the measurements $m_{i,j}$ are in watts, but they
may also be in monetary units, if the meter i knows the current unit price.

2.2.2   Security Model

This section presents Shannon's security model [10] re-written in the context of
smart grids. In this model, the meter i encrypts its measurement $m_{i,j}$, computes

$$\mathfrak{M}_{i,j} = \text{Enc}\left(m_{i,j}\right),$$

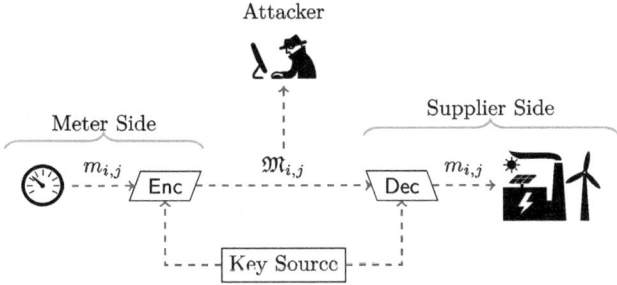

Fig. 2.5 Shannon's security model in the context of smart grids

and sends the encrypted measurement $\mathfrak{M}_{i,j}$ to the supplier, which decrypts as

$$m_{i,j} = \mathsf{Dec}\left(\mathfrak{M}_{i,j}\right).$$

The security model is composed of attack model and trust model. The former defines the capabilities assumed for the attackers. The latter defines the trust relationship between meters and their supplier with the corresponding changes to the attack model.

2.2.2.1 Attack Model

The attacker is very limited and can access only the encrypted measurement $\mathfrak{M}_{i,j}$. Note that there is no difference if the cryptographic scheme is either symmetric or asymmetric. However, Shannon's security model was created 27 years before the introduction of asymmetric cryptography [3]. Figure 2.5 depicts the attack model.

In this model, the attacker knows how the functions Enc and Dec work. The model does not allow hiding these functions, because security through obscurity is considered harmful. Hence, the attacker only does not know the keys and the measurement $m_{i,j}$. The security lies in the keys, which can be generated to create certificates with security and privacy without trusted third party (TTP) [1]. This work does not consider side-channel attack, fault attack, etc. A secure source generates the keys, used as input for the encryption and decryption functions. This source is not a TTP but a function or protocol as given in [1].

2.2.2.2 Trust Model

Usually, the meter i and its supplier are considered trusted. Thus, the meter measures the consumption correctly, computes Enc correctly, signs the result correctly, and sends the signed encrypted measurement directly to its supplier. The communication channel transmits the message without interruption and the supplier computes Dec correctly. There is no collusion.

Even with all these restrictions in the trust model, the attacker could infer information about the consumption if the encrypted measurements had a bijection with the measurements. The attacker could infer the encrypted measurement of zero watts and deduce when the customer is at home. To avoid such attack, the cryptographic function should be probabilistic.

2.2.2.3 Considerations About the Cryptographic Functions

In contrast to the well-known cryptographic functions that have the same encrypted measurement for the same measurement, we also have probabilistic encryption schemes that enable different encrypted measurements $\mathfrak{M}_{i,j}$ for the same measurement $m_{i,j}$. This is possible because probabilistic encryption schemes are based on additional parameters chosen by the meter. Such parameters are not necessary for the decryption function. Paillier cryptosystem [9] is an example of probabilistic encryption. In fact, if the meter i has the key k and a secret r, the encryption function should be written as

$$\mathfrak{M}_{i,j} = \mathsf{Enc}_{k,r}(m_{i,j}),$$

and the decryption function depends on the key $\bar{k}$ associated with k, thus the decryption function should be written as

$$m_{i,j} = \mathsf{Dec}_{\bar{k}}(\mathfrak{M}_{i,j}).$$

Moreover, if we have two secrets r_1 and r_2 such that $r_1 \neq r_2$, then

$$\mathsf{Enc}_{k,r_1}(m_{i,j}) \neq \mathsf{Enc}_{k,r_2}(m_{i,j}).$$

However,

$$\mathsf{Dec}_{\bar{k}}\left(\mathsf{Enc}_{k,r_1}(m_{i,j}) \odot \mathsf{Enc}_{k,r_2}(m_{i,j})\right) = m_{i,j} \oplus m_{i,j} = 2m_{i,j}, \qquad (2.1)$$

for all $m_{i,j}$. Section 6.4.1 uses this property to show that additive homomorphic encryption primitives (AHEPs) are particular cases of Asymmetric DC-Nets (ADC-Nets). Note that the functions form a bijection between two groups. Equation (2.1) denotes the operation over the measurements $m_{i,j}$ and the encrypted measurements $\mathfrak{M}_{i,j}$ as $\oplus$ and $\odot$, respectively. Note that encryption and decryption functions of probabilistic encryption schemes are usually presented without the keys neither the random number.

According to Shannon's terminology, the encrypted measurement is called ciphertext and the measurement is called message. In this work, message has different concepts. Message may refer to other packets sent in the information network.

Once the security is ensured in a system, we can go to the next challenge.

2.2.3 Privacy Model

Ensuring privacy is more complicated than ensuring security. The privacy model works under the assumption that the security model and its components are robust, i.e., if a security assumption fails, privacy is impaired.

The privacy model can be constructed using two strategies, namely: pseudonyms and data aggregation. The latter is adopted in this book and is more efficient for smart grids than the former, cf. Sect. 3.1.2 or [2]. The former requires that the measurements are associated with pseudonyms and sent through an anonymity network. Note that pseudonyms should be randomly chosen and unlinkable with each other. In particular, cryptographically secure pseudorandom number generators (CSPRNGs) are already computationally expensive. The latter relies on the idea of a ballot box. In other words, each meter i encrypts its measurement $m_{i,j}$ and somehow the encrypted measurements $\mathfrak{M}_{i,j}$ from all meters i in the round j are aggregated generating an encrypted consolidated consumption $\mathfrak{C}_j$, s.t.

$$\mathfrak{C}_j = \prod_{i=1}^{\tilde{\imath}} \mathfrak{M}_{i,j} = \prod_{i=1}^{\tilde{\imath}} \mathsf{Enc}\left(m_{i,j}\right).$$

After the aggregation, the supplier decrypts the encrypted consolidated consumption resulting in the consolidated consumption c_j, s.t.

$$c_j = \mathsf{Dec}\left(\mathfrak{C}_j\right) = \sum_{i=1}^{\tilde{\imath}} m_{i,j}.$$

2.2.3.1 Attack Model

The attacker is more powerful in a privacy attack model than in a security attack model. The attacker has access to the encrypted consolidated consumptions $\mathfrak{C}_j$ and all information on the supplier side, including the cryptographic key to decrypt them. The key source is still secure and distributes the keys to the meters and the supplier, which can decrypt only the encrypted consolidated consumption $\mathfrak{C}_j$ or even an individual measurement $m_{i,j}$, depending on the PPP used and if the supplier receives such measurement. Since AHEPs enable the decryption of a single measurement $m_{i,j}$, the attacker cannot have access to an individual encrypted measurement $\mathfrak{M}_{i,j}$, if the PPP is based on an AHEP. Figure 2.6 depicts a model for privacy and its data aggregation in the context of smart grids. Figure 2.6 does not have edges indicating the bill $b_i^{\$}$. The supplier already knows $b_i^{\$}$ for each meter i in a non-smart grid. More information about bills is presented further.

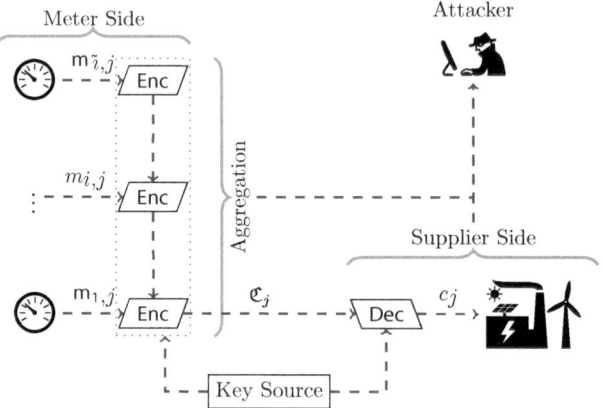

Fig. 2.6 Privacy model in the context of smart grids

2.2.3.2   Trust Model

In privacy trust models, we can define meters as trusted, honest-but-curious, or malicious. The first definition requires that meters behave correctly. This is a strong assumption, because machines might fail. The second requires that meters also behave correctly, but it will collect accessible information. In the honest-but-curious model, the attacker has no access to the communication channel during the aggregation process. However, it is clear that access to encrypted measurements $\mathfrak{M}_{i,j}$ should be denied for PPPs based on AHEPs. The third requires that meters might behave as an attacker. This is a safe, secure, and weak assumption, because factual or intentional failures can happen in real life. Parallel with these definitions, we could define meters as non-attackers, passive attackers, and active attackers, respectively.

For the privacy trust model, the supplier is malicious. This is a safe assumption for customers and even for the supplier, which do not need blindly to trust the employees.

Trusted meters measure the consumption correctly, compute Enc correctly, sign the result correctly, and send the signed encrypted measurements directly to their supplier. They do everything correctly.

Honest-but-curious also known as semi-honest meters behave like trusted meters, but they read information in the aggregation, if possible.

Malicious meters can fail to measure the correct consumption, can compute Enc wrongly, sign the result wrongly and send the signed encrypted measurements to their supplier and an attacker. The communication channel can transmit the messages with noise and interruption, and the supplier can compute Dec wrongly. Collusion is considered. Thus, an attacker has more information.

In contrast to previous work, this work presents PPPs taking in consideration that the meters might be malicious. Moreover, each meter and its supplier can verify the bill $b_i^{\$}$.

References

1. F. Borges, L.A. Martucci, M. Mühlhäuser, Analysis of privacy-enhancing protocols based on anonymity networks, in *2012 IEEE Third International Conference on Smart Grid Communications (SmartGridComm)* (2012), pp. 378–383. doi:10.1109/SmartGrid-Comm.2012.6486013
2. F. Borges et al., Secure and privacy-friendly public key generation and certification, in *2014 IEEE 13th International Conference on Trust, Security and Privacy in Computing and Communications (TrustCom)* (2014), pp. 114–121. doi:10.1109/TrustCom.2014.19
3. W. Diffie, M.E. Hellman, New directions in cryptography. IEEE Trans. Inform. Theory **22**(6), 644–654 (1976). issn:0018-9448. doi:10.1109/TIT.1976.1055638
4. P. Fox-Penner, *Smart Power: Climate Change, the Smart Grid, and the Future of Electric Utilities* (Island Press, Washington, DC, 2010). isbn:9781597268097
5. F. Katiraei, C. Sun, B. Enayati, No inverter left behind: protection, controls, and testing for high penetrations of pv inverters on distribution systems. IEEE Power Energ. Mag. **13**(2), 43–49 (2015). issn:1540-7977. doi:10.1109/MPE.2014.2380374
6. S. Li, K. Choi, K. Chae, An enhanced measurement transmission scheme for privacy protection in smart grid, in *2013 International Conference on Information Networking (ICOIN)* (2013), pp. 18–23. doi:10.1109/ICOIN.2013.6496345
7. D. Novosel, V. Rabl, J. Nelson, A report to the U.S. DOE: IEEE shares its insights on priority issues [leader's corner]. IEEE Power Energ. Mag. **13**(2), 6–12 (2015). issn:1540–7977. doi:10.1109/MPE.2014.2374971
8. T.G. Paraskevakos, *Sensor monitoring device*. US Patent 3,842,208 (1974). http://www.google.com/patents/US3842208
9. P. Paillier, Public-key cryptosystems based on composite degree residuosity classes, in *Advances in Cryptology - EUROCRYPT 1999*, vol. 1592. Lecture Notes in Computer Science (Springer, Berlin, 1999), pp. 223–238. isbn:978-3-540-65889-4
10. C.E. Shannon, Communication theory of secrecy systems. Bell Syst. Tech. J. **28**(4), 656–715 (1949). issn:0005-8580. doi:10.1002/j.1538-7305.1949.tb00928.x
11. J. Vasconcelos, Survey of regulatory and technological developments concerning smart metering in the European Union electricity market (2008). issn:1830-1541. http://hdl.handle.net/1814/9267

Chapter 3
A Selective Review

Abstract This chapter presents the areas in which Privacy-Preserving Protocols (PPPs) have been developed and aims to highlight the most relevant related work for PPPs. Naturally, there are privacy-enhancing technologies with restrictive results on cost, efficiency, or privacy. For example, the use of a home battery is the best solution as discussed in Sect. 3.1.1. However, it is too expensive. The areas with promising results are investigated in this book. The next two sections present the restrictive and promising results found.

Keywords Privacy-preserving protocols • Privacy-enhancing technologies • Survey • Obfuscation • Anonymization • Homomorphic encryption • DC-Net • Commitment

3.1 Solutions with Restrictive Result

This section presents interesting proposals in the literature that are not explored in the sequel of this book due to restrictions found. These solutions can be used to reduce the leakage of privacy [30]. For example, customers can use them to mask the signal patterns generated by their TVs. However, they cannot hide information from one day. For example, which of them is never at home on specific days. Even worse, these solutions cannot hide that some customers work in the middle of the night to achieve a goal before deadlines.

3.1.1 Data Obfuscation by Means of Storage Banks

Customers in smart grids formed by water suppliers can easily store water, and they can use it anytime without concern about privacy. Similarly, any kind of battery, i.e., energy storage system is good for the power load balance in a smart grid formed by energy suppliers, for instance, air-conditioning [18], water heaters, and electric vehicles [19] can be used as energy storage systems. Such batteries store the energy when the renewable sources have high electricity generation and discharge

© Springer International Publishing Switzerland 2017   25
F. Borges de Oliveira, *On Privacy-Preserving Protocols for Smart Metering Systems*,
DOI 10.1007/978-3-319-40718-0_3

when they have low generation. Any kind of energy storage system creates a buffer between supply and demand in a smart grid. Such a buffer protects privacy and boosts power load balancing. From the privacy point of view, non-intrusive load monitoring (NILM) [11] can analyze only the behavior of storages. In contrast to water storage, electricity storage is still too expensive to supply a house without the energy supplier for a day. Currently, energy storage is expensive even for energy suppliers, which prefer solutions based on curtailment or flexible generation [21].

When dwellers are at home, batteries from their electric vehicles can be used to protect their privacy. Besides electric vehicles bringing new privacy issues [25], not all dwellers will have an electric vehicle. Many customers might have one vehicle per family. They can buy small batteries to enhance their privacy [20, 28], but such solutions do not solve the problem. They reduce the problem only by means of creating a trade-off between the battery size and the leakage of privacy. Therefore, such a solution is still too expensive for the majority of customers.

3.1.2   Anonymization Via Pseudonymous

To avoid being profiled by their suppliers, each customer identity should be dismembered in at least two pseudonyms, one for high-frequency metering data and other for low-frequency metering data, in accordance with the nomenclature of Efthymiou and Kalogridis [10]. However, an attacker can easily relate bills to customers' measurements. Even though many bills have the same value, it is unlikely that many customers have the same bill in consecutive invoices. This drawback can be bypassed with the addition of new identities. The more identities, the more privacy. The best case for privacy is achieved with one identity per watt consumed, but this is the worst case for performance [3]. This trade-off suggests that other solutions are more interesting, because we search for efficient solutions that provide indistinguishability as additive homomorphic encryption primitives (AHEPs) do. Once attackers related the pseudonyms of a customer, they can read all measurements associated with the pseudonyms and apply NILM. Chapter 5 presents limitations for PPPs with aggregations. Such limitations are even worse for protocols based on pseudonyms, because the attacker knows the measurements—completely or partially—to link the customers with their measurements. This is known as de-anonymization. Chapter 5 clarifies how it is possible.

3.1.3   Data Obfuscation by Means of Noise Injection

This class of protocols adds noise to information for the attacker to receive scrambled data. A Gaussian or Laplacian distribution might insert noise [9]. Bohli et al. [1] present a solution based on expectation, i.e., each meter i adds its measurement $m_{i,j}$ to a random value $r_{i,j}$ generated by a known distribution with a known finite variance σ^2 and expectation μ. Thus, the encrypted consolidated consumption $\mathfrak{C}_j$ is given by

$$\mathfrak{C}_j = \sum_{i=1}^{\tilde{\imath}} m_{i,j} + r_{i,j} \approx \mu + \sum_{i=1}^{\tilde{\imath}} m_{i,j} \approx \mu + c_j.$$

Thus, if the supplier knows the distribution and the sum, it can compute the consolidated consumption c_j quickly. The expected value μ does not change per round j and can even be assumed to be zero, i.e., $\mu = 0$. Therefore, the meters can send their measurements directly to the supplier without extra communication and with very low processing time. A drawback found was the high number $\tilde{\imath}$ of meters necessary for the series to converge, i.e.,

$$\sum_{i=1}^{\tilde{\imath}} r_{i,j} \approx \mu.$$

However, Wang et al. [29] show that $\tilde{\imath}$ can be considerably smaller.

Addition of noise is not suitable for all privacy problems [9]. Smart grids present such a problem. Without loss of generality, suppose the measurements are collected every hour. Hence, we have 24 measurements per day. Thus, the average of the first measurement $m_{i,1}$ for the meter i is given by

$$m_{i,1} \approx \left(\sum_{l=0}^{\tilde{\imath}} \mathfrak{M}_{i,24l+1} \right) / \tilde{\imath}, \tag{3.1}$$

assuming that customers have routines, and therefore, the measurements in the same hour are close to their average. To find the second expected measurement $m_{i,2}$, one just changes the index $24l + 1$ in Eq. (3.1) to $24l + 2$, for the third, $24l + 3$, etc. Therefore, an attacker in the supplier side can create a profile with the expected value for the measurements for each customer. Later, the attacker can sell information about the customers' habits to health insurance companies, for instance.

3.2 Solutions Addressed in This Book: Anonymization Via Cryptographic Protocols

This section presents protocols that inspired the solutions proposed in this book. The understanding of these protocols simplifies the understanding of the first three protocols presented in Chap. 6. Sections 3.2.1 and 3.2.2 present protocols that provide only the consolidated consumption c_j, while Sect. 3.2.3 presents a protocol that provides only bill $b_i^\$$ with verification. However, there are also protocols in the literature that provide both c_j and $b_i^\$$ [2]. Moreover, there are protocols that fulfill the requirements in Sect. 4.2, e.g., [4, 5, 7].

This section only addresses the cryptographic protocols for anonymization used to improve the state of the art. Many other cryptographic schemes can be used to create anonymization, e.g., Shamir Secret Sharing [27] can be used to provide anonymity in smart grid scenarios [24].

3.2.1   Protocols Based on Homomorphic Encryption

An AHEP is a cryptographic algorithm based on a function with the property

$$\prod_{i=1}^{\tilde{i}} \mathsf{Enc}\left(m_{i,j}\right) = \mathsf{Enc}\left(\sum_{i=1}^{\tilde{i}} m_{i,j}\right).$$

If the PPP requires only this property, it can use any AHEP. This is the case for [14, 17, 26]. Due to performance reasons, they chose the Paillier cryptosystem [22], which is used in many protocols for smart grid scenarios [2]. His encryption function $\mathsf{Enc}\left(m_{i,j}\right)$ of the measurement $m_{i,j}$ is given by

$$\begin{aligned} \mathsf{Enc} &: \mathbb{Z}_n \times \mathbb{Z}_n^* \to \mathbb{Z}_{n^2} \\ \mathsf{Enc}(m_{i,j}, r_{i,j}) &\mapsto g^{m_{i,j}} \cdot r_{i,j}^n \mod n^2, \end{aligned} \tag{3.2}$$

where n is the product of two safe primes p and q, and g and $r_{i,j}$ are random numbers chosen by the supplier and meter, respectively. To ensure bijectivity, the n should divide the order of $g \in \mathbb{Z}_{n^2}^*$.

Paillier is an AHEP over $\mathbb{Z}_n$ with the product of the encrypted measurements over $\mathbb{Z}_{n^2}$. For example, the consolidated consumption is given by

$$\begin{aligned} \mathfrak{C}_j &= \mathsf{Enc}\left(m_{1,j}, r_{1,j}\right) \cdots \mathsf{Enc}\left(m_{\tilde{i},j}, r_{\tilde{i},j}\right) = \prod_{i=1}^{\tilde{i}} \mathsf{Enc}\left(m_{i,j}, r_{i,j}\right) \\ &= g^{m_{1,j}} r_{1,j}^n \cdots g^{m_{\tilde{i},j}} \cdot r_{\tilde{i},j}^n \mod n^2 \\ &= g^{\sum_{i=1}^{\tilde{i}} m_{i,j}} \prod_{i=1}^{\tilde{i}} r_{i,j}^n \mod n^2 \\ &= \mathsf{Enc}\left(\sum_{i=1}^{\tilde{i}} m_{i,j}, r_{i,j}, \prod_{i=1}^{\tilde{i}} r_{i,j}\right). \end{aligned} \tag{3.3}$$

Algorithm 1: Paillier

---

1 ProcedureEncryption

　　Input: measurements $m_{i,j}$.

　　Output: encrypted measurements $\mathfrak{M}_{i,j}$.

2　　for $i \leftarrow 1$ to $\tilde{\imath}$ do

3　　　$\lfloor$　$\mathfrak{M}_{i,j} \leftarrow \mathsf{Enc}\,(m_{i,j})$ // v.s. Eq. (3.2)

4 Procedure Aggregation

　　Input: encrypted measurements $\mathfrak{M}_{i,j}$.

　　Output: encrypted consolidated consumption $\mathfrak{C}_j$.

5　　$\mathfrak{C}_j \leftarrow 1$

6　　for $i \leftarrow 1$ to $\tilde{\imath}$ do

7　　　$\lfloor$　$\mathfrak{C}_j \leftarrow \mathfrak{C}_j \cdot \mathfrak{M}_{i,j}$ // v.s. Eq. (3.3)

8 Procedure Decryption

　　Input: encrypted consolidated consumption $\mathfrak{C}_j$.

　　Output: consolidated consumption c_j.

9　　$\lfloor$ $c_j \leftarrow \mathsf{Dec}\,(\mathfrak{C}_j)$ // v.s. Eq. (3.4)

The public key is given by $\{n, g\}$ and the private key is given by $d = \mathrm{L}(g^{\lambda}$ mod $n^2)^{-1}$ defined by Carmichael's function $\lambda = \lambda(n) = \mathsf{lcm}\,(p-1, q-1)$, where lcm is the function that returns the least common multiple. His decryption function Dec is given by

$$\mathsf{Dec} : \mathbb{Z}_{n^2} \to \mathbb{Z}_n$$
$$\mathsf{Dec}(\mathfrak{C}_j) \mapsto \mathrm{L}(\mathfrak{C}_j^{\lambda} \mod n^2) \cdot d \mod n, \tag{3.4}$$

where $\mathrm{L}(u) = (u-1)/n$.

Probabilistic encryption is one requirement for AHEPs, because the encryption function of measurements with the same value should return different encrypted measurements. Therefore, the decryption function does not depend on the random numbers, and therefore, we can write

$$\mathsf{Dec}\left(\prod_{i=1}^{\tilde{\imath}} \mathsf{Enc}\,(m_{i,j}) \mod n^2\right) = \sum_{i=1}^{\tilde{\imath}} m_{i,j} \mod n$$

without the random number. Algorithm 1 describes the procedures in the Paillier cryptosystem.

Since the supplier has the private key, it can decrypt a single measurement. Thus, it should receive only the encrypted consolidated consumption. Hence, schemes based on AHEPs need a trusted aggregator, which might be a trusted third party (TTP) or operations between the meters and their supplier, i.e., a virtual aggregator. Figure 3.1 depicts a communication model for schemes based on AHEPs.

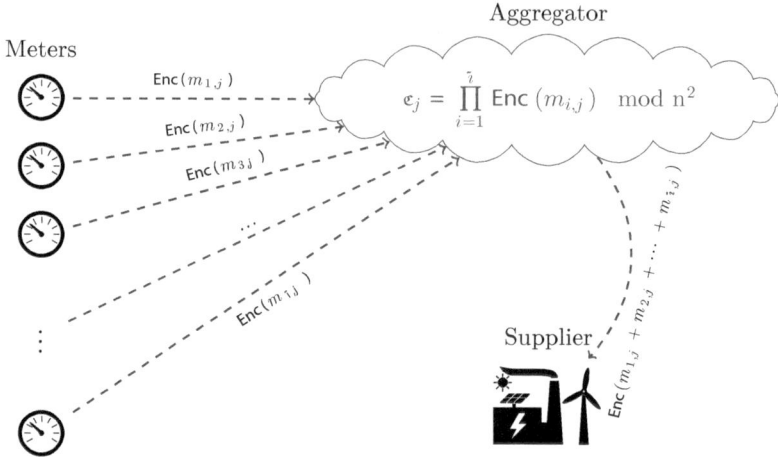

Fig. 3.1 Abstraction of an aggregator for AHEPs

In [17], Li et al. present a PPP avoiding a TTP with a technique called in-network aggregation, i.e., the meters send their measurements to each other until they perform the aggregation. The last meter sends an encrypted consolidated consumption $\mathfrak{C}_j$ to the supplier. This technique should assume the meters are honest-but-curious and an attacker cannot spoof their communication. In [26], Ruj and Nayak present another kind of virtual aggregator. They use access control to compute the aggregation in the network devices. However, they need the same assumptions.

3.2.2   Protocols Based on DC-Nets

In [8], Chaum introduces the DC-Net protocol to provide anonymous communication. The name comes from Dining Cryptographers. They want to discover if one of them paid for the dinner, but they do not want to reveal the identity of who paid. The DC-Net protocol is symmetric, thus the number of keys grows quadratically with respect to the number of users.

In [12], Erkin and Tsudik present a DC-Net that provides consolidated consumption c_j using the Paillier Cryptosystem. In [16], Kursawe et al. present many ways of using DC-Nets resulting in the consolidated consumption c_j. The most efficient way is called Low-Overhead Protocol (LOP) and is presented in this section. Let us call the DC-Nets introduced by Chaum as Symmetric DC-Nets (SDC-Nets) to differentiate them from Asymmetric DC-Nets (ADC-Nets) [6]. The core difference between a fully connected SDC-Net and the LOP is in the number of bits for integer representation. LOP uses only integers with 32 bits.

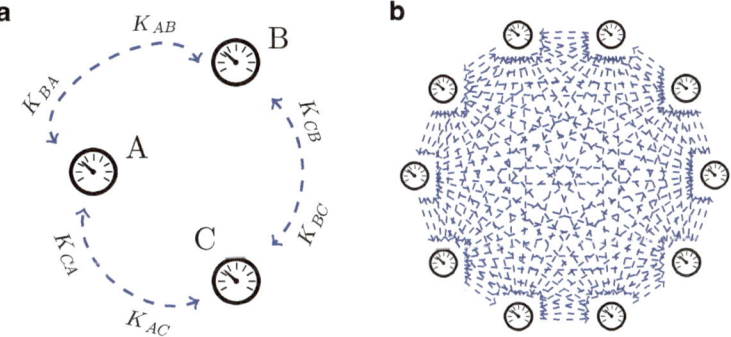

Fig. 3.2 Key exchange of fully connected SDC-Nets. (**a**) With three meters. (**b**) With ten meters

In the set-up phase, a set $\mathcal{M}$ of meters agrees on a secure hash function s.t. it behaves as a one-way function and has collision resistance. They send a symmetric key to each other, i.e., $k_{i \to o}$. Figure 3.2 depicts the key exchange between meters in SDC-Nets. Without loss of generality, Fig. 3.2a depicts the key exchange of three meters, and Fig. 3.2b depicts the case for ten meters, where we can see that the process is not well scalable.

Thereafter, they can encrypt computing

$$\mathfrak{M}_{i,j} \overset{\text{def.}}{=} \mathsf{Enc}\left(m_{i,j}\right) \overset{\text{let}}{=} m_{i,j} + \sum_{o \in \mathcal{M}-\{i\}} (-1)^{o<i} \, \mathsf{H}\left(k_{i \to o} \,||\, j\right), \tag{3.5}$$

where $||$ denotes string concatenation and H is the hash function.

The aggregation happens together with the description, namely,

$$c_j \overset{\text{def.}}{=} \mathsf{Dec}\left(\{\mathfrak{M}_{i,j} | i \in \mathcal{M}\}\right) \overset{\text{let}}{=} \sum_{i=1}^{\tilde{\imath}} \mathfrak{M}_{i,j}. \tag{3.6}$$

Algorithm 2 describes the procedures to this protocol—LOP for integers with 32 bits—while Fig. 3.3 depicts the communication network with three meters, without loss of generality.

Note that Eq. (3.5) can be seen as

$$\mathfrak{M}_{i,j} \overset{\text{def.}}{=} \mathsf{Enc}\left(m_{i,j}\right) = m_{i,j} + \sum_{o \in \mathcal{M}-\{i\}} \mathsf{H}\left(k_{i,o} \,||\, j\right) - \mathsf{H}\left(k_{o,i} \,||\, j\right). \tag{3.7}$$

Equation (3.7) can be easily related to Fig. 3.3 and Algorithm 2.

Algorithm 2: LOP—SDC-Net for 32 bits

---

1 **Procedure** Encryption

    Input: measurements $m_{i,j}$.

    Output: encrypted measurements $\mathfrak{M}_{i,j}$.

2   **for** $i \leftarrow 1$ **to** $\tilde{i}$ **do**

3          $\mathfrak{M}_{i,j} \leftarrow m_{i,j}$

4   **for** $o \leftarrow 1$ **to** $\tilde{i}$ **do**

5   **if** $i \neq o$ **then**

6                  $\mathfrak{M}_{i,j} \leftarrow \mathfrak{M}_{i,j} + \mathsf{H}\left(k_{i,o}\|j\right) - \mathsf{H}\left(k_{o,i}\|j\right)$

7 **Procedure** Decryption

    Input: encrypted measurements $\mathfrak{M}_{i,j}$.

    Output: consolidated consumption c_j.

8      $c_j \leftarrow 0$

9   **for** $i \leftarrow 1$ **to** $\tilde{i}$ **do**

10          $c_j \leftarrow c_j + \mathfrak{M}_{i,j}$

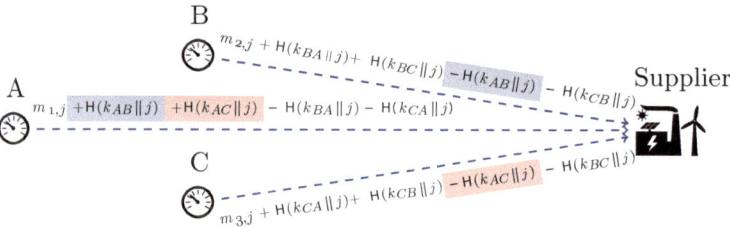

Fig. 3.3 A fully connected SDC-Net

3.2.3 Protocols Based on Commitment

In [15], Jawurek et al. present a PPP based on Pedersen Commitments [23] to enable billing verification, i.e., the supplier can verify whether a meter i sent the correct bill $b_i^\$$. The protocol requires a plug-in privacy component PC_i attached to the meter i. The PC_i is a hardware that receives from the meter i the measurements $m_{i,j}$, random values $r_{i,j}$, commitments $\mathfrak{N}_{i,j}$, and digital signature $\mathfrak{S}_{i,j}$ for the commitments, which are calculated with Pedersen Commitments as follows:

$$\mathfrak{N}_{i,j} \stackrel{\text{def.}}{=} \mathsf{Commit}\left(m_{i,j}, r_{i,j}\right) \stackrel{\text{let}}{=} g^{m_{i,j}} \cdot h^{r_{i,j}}, \tag{3.8}$$

where g and h are random numbers belonging to $\mathbb{Z}_p^*$, i.e., a multiplicative group of integers $\mathbb{Z}$ modulo p, where p is prime. Concurrently, the PC_i receives the electricity tariff $T = (t_1, t_2, \ldots, t_{\tilde{j}})$ from the supplier and sends the signed commitments, P, and r' given by

$$b_i^\$ \stackrel{\text{def.}}{=} \sum_{j=1}^{\tilde{j}} m_{i,j} \cdot t_j$$

and

$$r' \stackrel{\text{def.}}{=} \sum_{j=1}^{\tilde{j}} r_{i,j} \cdot t_j.$$

To check the bill $b_i^\$$, the supplier generates a committed consolidated measurement $\mathfrak{T}_i$ computing

$$\mathfrak{T}_i \stackrel{\text{def.}}{=} \prod_{j=1}^{\tilde{j}} \mathfrak{N}_{i,j}^{t_j}.$$

The PC$_i$ declares the bill $b_i^\$$, and the supplier verifies whether $b_i^\$$ is correct, i.e., if the supplier can open the commitment

$$\mathsf{Open}\left(\mathfrak{T}_i, b_i^\$, r'\right) \stackrel{\text{let}}{=} \left(g^{b_i^\$} \cdot h^{r'} \stackrel{?}{=} g^{\sum_{i=1}^{\tilde{i}} m_{i,j} \cdot t_j} \cdot h^{\sum_{i=1}^{\tilde{i}} r_{i,j} \cdot t_j}\right). \tag{3.9}$$

The supplier can verify by the signatures whether the PC$_i$ computed the values correctly. Algorithm 3 describes the procedures of this protocol for each meter. Note that the PC$_i$ could be inside the meter i, and therefore, the measurements would never leave the meters. Figure 3.4 depicts the communication between a meter and its supplier. Note that meters work independently from each other in this protocol.

Researchers are discovering new applications and benefits of smart meters. Specifically, they are discovering new applications that require consolidated consumptions. Billing does not require consolidated consumptions, and therefore, the smart meters do not need to send encrypted measurements for billing. Moreover, no one application justifies the suppliers to receive individual measurements, and the majority of the PPPs address only the aggregation for the consolidated consumption [2, 4, 7].

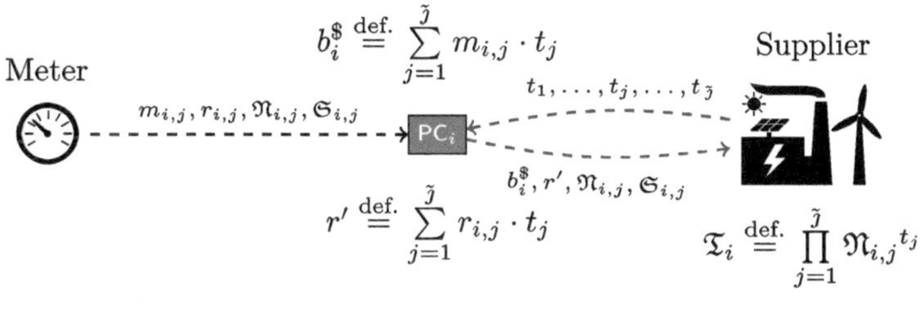

Fig. 3.4 Communication model for the billing verification

Algorithm 3: Billing verification

---

1 Procedure Meter

    Output: measurements $m_{i,j}$, random numbers $r_{i,j}$, commitments $\mathfrak{N}_{i,j}$, and signatures $\mathfrak{S}_{i,j}$.

2   **for** $j \leftarrow 1$ **to** $\tilde{j}$ **do**

3 $r_{i,j} \leftarrow$ Random number

4 $\mathfrak{N}_{i,j} \leftarrow$ Commit $\left(m_{i,j}, r_{i,j}\right)$ // v.s. Eq. (3.8)

5 $\mathfrak{S}_{i,j} \leftarrow$ Sign $\left(\mathfrak{N}_{i,j}\right)$

6 Procedure PC_i

    Input: measurements $m_{i,j}$, random numbers $r_{i,j}$, commitments $\mathfrak{N}_{i,j}$, signatures $\mathfrak{S}_{i,j}$, the tariffs t_j.

    Output: commitments $\mathfrak{N}_{i,j}$, signatures $\mathfrak{S}_{i,j}$, and bill $b_i^{\$}$ in monetary value.

7 $b_i^{\$} \leftarrow 0$

8 $r' \leftarrow 0$

9   **for** $j \leftarrow 1$ **to** $\tilde{j}$ **do**

10 $b_i^{\$} \leftarrow b_i^{\$} + m_{i,j} \cdot t_j$

11 $r' \leftarrow r' + r_{i,j} \cdot t_j$

12 Procedure Supplier

    Input: commitments $\mathfrak{N}_{i,j}$, signatures $\mathfrak{S}_{i,j}$, and bill $b_i^{\$}$ in monetary value.

13 $\mathfrak{N}_{i,j} \leftarrow 1$

14   **for** $j \leftarrow 1$ **to** $\tilde{j}$ **do**

15   **if** Verify $\left(\mathfrak{N}_{i,j}, \mathfrak{S}_{i,j}\right)$ **then**

16 $\mathfrak{N}_{i,j} \leftarrow \mathfrak{N}_{i,j} \cdot \mathfrak{N}_{i,j}^{t_j}$

17   **else**

18   Apply policies

19   **if** Open $\left(\mathfrak{T}_i, b_i^{\$}, r'\right)$ // v.s. Eq. (3.9)

20   **then**

21   Bill is correct

22   **else**

23   Apply policies

References

1. J.-M. Bohli, C. Sorge, O. Ugus, A privacy model for smart metering, in *2010 IEEE International Conference on Communications Workshops (ICC)* (2010), pp. 1–5. doi:10.1109/ICCW.2010.5503916

2. F. Borges, L.A. Martucci, iKUP keeps users' privacy in the Smart Grid, in *2014 IEEE Conference on Communications and Network Security (CNS)* (2014), pp. 310–318. doi:10.1109/CNS.2014.6997499

3. F. Borges, M. Mühlhäuser, EPPP4SMS: efficient privacy-preserving protocol for smart metering systems and its simulation using real-world data. IEEE Trans. Smart Grid **5**(6), 2701–2708 (2014). doi:10.1109/TSG.2014.2336265. http://dx.doi.org/10.1109/TSG.2014.2336265

4. F. Borges, L.A. Martucci, M. Mühlhäuser, Analysis of privacy-enhancing protocols based on anonymity networks, in *2012 IEEE Third International Conference on Smart Grid Communications (SmartGridComm)* (2012), pp. 378–383. doi:10.1109/SmartGrid-Comm.2012.6486013

5. F. Borges et al., A privacy-enhancing protocol that provides innetwork data aggregation and verifiable smart meter billing, in *2014 IEEE Symposium on Computers and Communication (ISCC)* (2014), pp. 1–6. doi:10.1109/ISCC.2014.6912612

6. F. Borges, J. Buchmann, M. Mühlhäuser, Introducing asymmetric DC-Nets, in *2014 IEEE Conference on Communications and Network Security (CNS)* (2014), pp. 508–509. doi:10.1109/CNS.2014.6997528

7. F. Borges, F. Volk, M. Mühlhäuser, Efficient, verifiable, secure, and privacy-friendly computations for the smart grid, in *2015 IEEE Power Energy Society Innovative Smart Grid Technologies Conference (ISGT)* (2015), pp. 1–5. doi:10.1109/ISGT.2015.7131862

8. D. Chaum, The dining cryptographers problem: unconditional sender and recipient untraceability. J. Cryptol. **1**(1), 65–75 (1988). issn:0933-2790. http://dl.acm.org/citation.cfm?id=54235.54239

9. C. Dwork, Differential privacy: a survey of results. English, in *Theory and Applications of Models of Computation*, ed. by M. Agrawal et al., vol. 4978. Lecture Notes in Computer Science (Springer, Berlin, Heidelberg, 2008), pp. 1 –19. isbn:978-3-540-79227-7. doi:10.1007/978-3-540-79228-4_1. http://dx.doi.org/10.1007/978-3-540-79228-4_1

10. C. Efthymiou, G. Kalogridis, Smart grid privacy via anonymization of smart metering data, in *2010 First IEEE International Conference on Smart Grid Communications (SmartGridComm)* (2010), pp. 238–243. doi:10.1109/SMARTGRID.2010.5622050

11. G. Eibl, D. Engel, Influence of data granularity on smart meter privacy. IEEE Trans. Smart Grid **6**(2), 930–939 (2015). issn:1949-3053. doi:10.1109/TSG.2014.2376613

12. Z. Erkin, G. Tsudik, Private computation of spatial and temporal power consumption with smart meters, in *ACNS*, ed. by F. Bao, P. Samarati, J. Zhou, vol. 7341. Lecture Notes in Computer Science (Springer, Berlin, 2012), pp. 561–577. isbn:978-3-642-31283-0

13. Z. Erkin et al., Privacy-preserving data aggregation in smart metering systems: an overview. IEEE Signal Process. Mag. 30.2 (2013), pp. 75–86. issn: 1053-5888. doi:10.1109/MSP.2012.2228343.

14. M. Jawurek, F. Kerschbaum, Fault-tolerant privacy- preserving statistics, in *Privacy Enhancing Technologies*, ed. by S. Fischer-Hübner, M. Wright, vol. 7384. Lecture Notes in Computer Science (Springer, Berlin, Heidelberg, 2012), pp. 221–238. isbn:978-3-642-31679-1. doi:10.1007/978-3-642-31680-7_12. http://dx.doi.org/10.1007/978-3-642-31680-7_12

15. M. Jawurek, M. Johns, F. Kerschbaum, Plug-in privacy for smart metering billing, in *Proceedings of Privacy Enhancing Technologies: 11th International Symposium, PETS 2011, Waterloo, ON, July 27–29, 2011*. ed. by S. Fischer-Hübner, N. Hopper (Springer, Berlin, Heidelberg, 2011), pp. 192–210. isbn:978-3-642-22263-4. doi:10.1007/978-3-642-22263-4_11. http://dx.doi.org/10.1007/978-3-642-22263-4_11

16. K. Kursawe, G. Danezis, M. Kohlweiss, Privacy-friendly aggregation for the smart-grid, in *Proceedings of Privacy Enhancing Technologies: 11th International Symposium, PETS 2011, Waterloo, ON, July 27–29, 2011*, ed. by S. Fischer-Hübner, N. Hopper (Springer, Berlin, Heidelberg, 2011), pp. 175–191. isbn:978-3-642-22263-4. doi:10.1007/978-3-642-22263-4_10. http://dx.doi.org/10.1007/978-3-642-22263-4_10

17. F. Li, B. Luo, P. Liu, Secure information aggregation for smart grids using homomorphic encryption, in *2010 First IEEE International Conference on Smart Grid Communications (SmartGridComm)* (2010), pp. 327–332. doi:10.1109/SMARTGRID.2010.5622064

18. N. Lu, An Evaluation of the HVAC load potential for providing load balancing service. IEEE Trans. Smart Grid **3**(3), 1263–1270 (2012). issn:1949-3053. doi:10.1109/TSG.2012.2183649

19. T. Masuta, A. Yokoyama, Supplementary load frequency control by use of a number of both electric vehicles and heat pump water heaters. IEEE Trans. Smart Grid **3**(3), 1253–1262 (2012). issn:1949-3053. doi:10.1109/TSG.2012.2194746

20. S. McLaughlin, P. McDaniel, W. Aiello, Protecting consumer privacy from electric load monitoring, in *Proceedings of the 18th ACM Conference on Computer and Communications Security*. CCS '11. Chicago, IL (ACM, New York, 2011), pp. 87–98. isbn:978-1-4503-0948-6. doi: 10.1145/2046707.2046720. http://doi.acm.org/10.1145/2046707.2046720

21. D. Novosel, V. Rabl, J. Nelson, A report to the U.S. DOE: IEEE shares its insights on priority issues [leader's corner]. IEEE Power Energ. Mag. **13**(2), 6–12 (2015). issn:1540-7977. doi:10.1109/MPE.2014.2374971

22. P. Paillier, Public-key cryptosystems based on composite degree residuosity classes, in *Advances in Cryptology - EUROCRYPT 1999*, vol. 1592. Lecture Notes in Computer Science (Springer, Berlin, 1999), pp. 223–238. isbn:978-3-540-65889-4

23. T.P. Pedersen, Non-interactive and information-theoretic secure verifiable secret sharing, in *Proceedings of the 11th Annual International Cryptology Conference on Advances in Cryptology*. CRYPTO '91 (Springer, London, 1992), pp. 129–140. isbn:3-540-55188-3. http://dl.acm.org/citation.cfm?id=646756.705507.

24. C. Rottondi, G. Verticale, C. Krauss, Distributed privacy-preserving aggregation of metering data in smart grids. IEEE J. Sel. Areas Commun. **31**(7), 1342–1354 (2013). issn:0733-8716. doi:10.1109/JSAC.2013.130716

25. C. Rottondi, S. Fontana, G. Verticale, A privacy-friendly framework for vehicle-to-grid interactions. English, in *Smart Grid Security*, ed. by J. Cuellar. Lecture Notes in Computer Science (Springer International Publishing, Cham, 2014), pp. 125–138. isbn:978-3-319-10328-0. doi:10.1007/978-3-319-10329-7_8. http://dx.doi.org/10.1007/978-3-319-10329-7_8.

26. S. Ruj, A. Nayak, A decentralized security framework for data aggregation and access control in smart grids. IEEE Trans. Smart Grid **4**(1), 196–205 (2013). issn:1949-3053. doi:10.1109/TSG.2012.2224389

27. A. Shamir, How to share a secret. Commun. ACM **22**(11), 612–613 (1979). issn:0001-0782. doi:10.1145/359168.359176. http://doi.acm.org/10.1145/359168.359176

28. D. Varodayan, A. Khisti, Smart meter privacy using a rechargeable battery: minimizing the rate of information leakage, in *2011 IEEE International Conference on Acoustics, Speech and Signal Processing (ICASSP)* (2011), pp. 1932–1935. doi:10.1109/ICASSP.2011.5946886

29. S. Wang et al., A randomized response model for privacy preserving smart metering. IEEE Trans. Smart Grid **3**(3), 1317–1324 (2012). issn:1949-3053. doi:10.1109/TSG.2012.2192487

30. W. Yang et al., Minimizing private data disclosures in the smart grid, in *Proceedings of the 2012 ACM Conference on Computer and Communications Security*. CCS '12. (ACM, Raleigh, NC, 2012), pp. 415–427. isbn:978-1-4503-1651-4. doi:10.1145/2382196.2382242. http://doi.acm.org/10.1145/2382196.2382242

Part II
Contributions

"Aha! So, that's your secret informant, a machine!" Said the post-doc. "There is no spy! You used the chronoscope to reveal the national top-secret plan among other things. Didn't you?"

Thinking of the consequences, the professor felt a consternation and subtly frowned.

"It's amazing! Now, we can avoid crimes, we can know who is guilty or innocent. There will be no more wrongdoing. We just need to adjust the chronoscope, which can measure every flash of light, every drop of water, every warm up, every cool off, etc. The data is always available and the most important moments of a lifetime can be quickly caught and played. When you were there in silence without moving in the darkness, even though, chronoscope can disclose the period that you were there and if you were alone." The astonishment was becoming anxiety. "It cannot read your mind but can reveal what you like to see, to watch, to listen, to make, etc. With complete information, efficient algorithms can predict your actions. It doesn't matter what you think. Actions speak louder than words." It is easy to see that concern was the unique felling.

After a short period of silence, the monologue continued. "If we let people know about it, they will start watching each other. Parents and teenagers, employees and employers, wives and husbands, competitors, adversaries, each one keeping a close watch on the other instead of doing the duties. We need to destroy the chronoscope... but someone else can rediscover it, manipulate individuals, and the whole society. Now, a computer can explain, predict, and control all behavior. It should be regulated and supervised..."

More silence, anyone could be watching.

For millenniums, each one could be alone, cry in the silence, come away, be anonymous, be away from intolerant humans, and learn by trial and error. They could acquire familiarity, intimacy, and complicity. They had moments of privacy until the chronoscope. The human being had developed several beliefs and religions, which preached a benevolent, omnipotent, and omniscient God. However, they clearly distinguished between the intangible benevolent and prying watchers. Previously, each habit, each ritual, each tradition, each entertainment, each choice, each exchanged glance, and each tiniest act kept a certain amount of privacy.

After a few minutes with the chronoscope, a mild paranoiac thought came over as "...too many secrets' the world will never be the same..."

—Inspired by Isaac Asimov, *The Dead Past*

Chapter 4
Reasons to Measure Frequently and Their Requirements

Abstract This chapter presents three reasons for smart meters to measure frequently and their requirements for smart grid scenarios. Privacy-Preserving Protocols (PPPs) based on commitment functions can provide billing with dynamic pricing without measurements leaving the smart meters. Thus, these reasons require information about the measurements on the supplier side.

Keywords Measurements • Smart grid • Fraud • Energy loss • Virtualization • Distribution • Minimum requirements • Consumption

The National Institute of Standards and Technology (NIST)[1] presents the advanced metering infrastructure (AMI) as a key mechanism to achieve dynamic pricing and *demand response*, which are necessary to match generation and consumption, creating an electric load balancing. Normally, meters are associated with time-based pricing [2] and with frequent measurements that are intrusive, but these two associations are independent of each other. Time-based pricing can be achieved without the frequent measurements leaving the meters [3], cf. Sect. 3.2.3. For forecasting, a phasor measurement unit (PMU) provides more information than the consolidated consumption provided by the meters. It provides information about the electricity quality. A PMU can measure aggregated measurements in a cell of the power network. Aggregated measurements a_j can also be achieved with smart meters spreading over the power network. Figure 4.1 depicts a smart grid for an energy supplier with the power network and the information network where the PMU provides aggregated measurements a_j to the energy supplier forecasting the energy consumption, and the meters sending their billed consumptions b_i to their energy supplier weekly or monthly. At this point, we are wondering if there is a reason for meters to send their measurements or their encrypted measurements to their energy supplier. This chapter presents three reasons raised by the author.

---

[1]Publication 1108R2.

© Springer International Publishing Switzerland 2017

F. Borges de Oliveira, *On Privacy-Preserving Protocols for Smart Metering Systems*, DOI 10.1007/978-3-319-40718-0_4

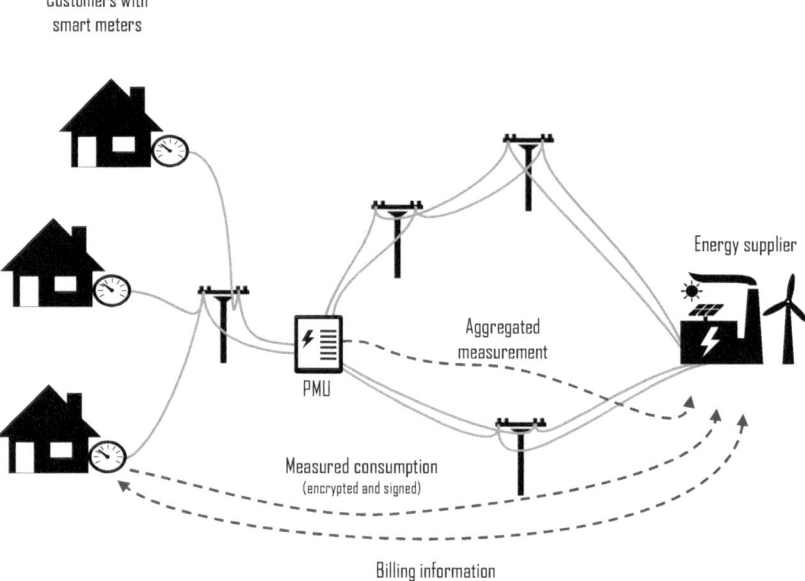

Fig. 4.1 Scheme of a smart grid

4.1 Reasons for Frequent Measurements

A previous work [5] presents that measurements sent to the supplier can be used
to detect overload in old distribution transformers and to protect them. This book
presents three new reasons for meters to send their frequent measurements to their
supplier, namely, to improve detection of fraud and energy loss, virtualization of the
power network, and fair distribution. Indeed, the energy supplier needs information
about the measurements $m_{i,j}$ to compute the consolidated consumptions c_j. In addition, Chap. 6 presents protocols that compute c_j from the encrypted measurements
$\mathfrak{M}_{i,j}$. Therefore, the energy supplier only needs the encrypted measurements $\mathfrak{M}_{i,j}$.

4.1.1 Fraud and Energy Loss

Fraud has been a problem for energy suppliers [7]. If a customer bypasses a
meter, the supplier suffers a fraud with energy loss. Time-based pricing opens more
opportunities for fraud because the amount of electricity consumed—billed consumption b_i—can be correct, but its monetary value—bill $b_i^\$$—may not. However,
fraud and energy loss might also be independent. Fraud might be financial without
energy loss and energy loss might be accidental. Without receiving the consolidated
consumption c_j, the energy supplier can detect only energy loss between the

PMU and meters after comparing the aggregated measurements a_j with the billed consumptions b_i, which is currently collected either on a monthly or on a yearly basis in the majority of countries. In other words, the supplier can verify only whether

$$\sum_i b_i \stackrel{?}{=} \sum_j a_j - \epsilon_j$$

and

$$\sum_i b_i^\$ \stackrel{?}{=} \sum_j \text{Value}\left(a_j - \epsilon_j\right)$$

hold either monthly or yearly, where the function Value returns the monetary value of the aggregated measurements, and ϵ_j is the transmission cost in the round j. Certainly, the consolidated consumption is different from the aggregated measurement, i.e., $c_j \neq a_j$ due to the transmission cost, but the values should be close $c_j \approx a_j$ and different by constants ϵ_j, s.t. $c_j = a_j - \epsilon_j$ and the values of the sequence ϵ_j are close to each other. Note that ϵ_j depends on the resistance of electrical equipment in the power network, for instance cables and transformers.

A month or a year is a lot of time to detect if something is going wrong. With the consolidated consumption, the supplier just verifies whether $c_j \approx a_j$ is a good approximation. Thus, the supplier can detect any sort of fraud and energy loss between a PMU and the meters. The idea relies on the assumption that the electric current that passes through meters also passes through a unique PMU, i.e., it is in between a set of meters and their supplier. Clustering the meters in disjointed sets might sound strange in a highly connected power network. However, for equivalent systems, the connection of power sources in series provides higher voltage and in parallel provides higher current. Whereas the power network has standards ensuring that all meters should receive the same power with constant voltage, the supplier should install a transformer for the set of meters with more than one power source. Therefore, a PMU can be installed and the meters can be clustered in disjointed sets.

Note that other kinds of suppliers have similar problems with fraud and loss of commodities. The loss of other commodities might be even worse than electricity, for instance, water leaking in the pipes causing infiltration. If no one detects the infiltration after a long time, it might cause erosion or even a sinkhole.

4.1.2 Virtualization of the Supplier Commodity Network

The virtualization of the supplier commodity network is the creation of multiple commodity networks over the same physical infrastructure. Thus, multiple suppliers can share the same infrastructure to distribute their commodities between their customers. Specifically, each supplier supplies the commodity network with the

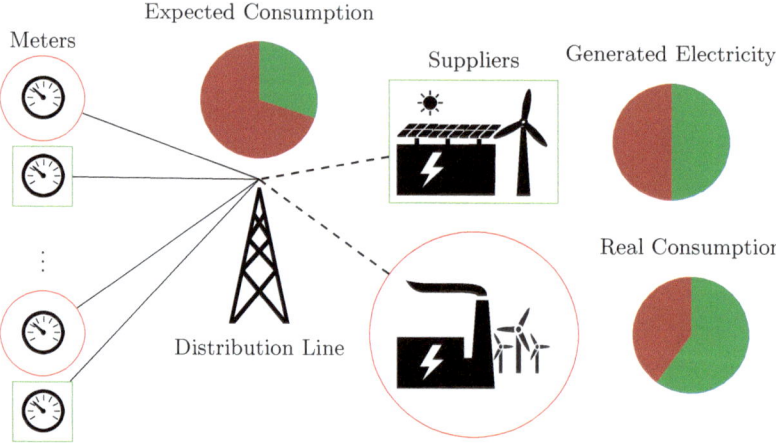

Fig. 4.2 Multiple suppliers needing frequent consolidated consumption c_j

amount necessary for its customers, and then, the commodity is fairly distributed between the customers of all suppliers. Certainly, the commodity from different suppliers should have the same quality.

Let us keep the focus on a power network because the electricity price tends to be more unstable and even can become negative [6], because wind farms and solar panels are unstable sources of energy, adding inconsistency in the pricing. Thus, consider a distribution network with a small number of meters, so small that it has only one PMU. However, two energy suppliers share the power network. They should provide the correct amount of electricity that their respective customers are consuming. A little bit more or less electricity is too much or too little. If the amount of electricity crosses some thresholds, the power network has a power outage. In this scenario, each energy supplier should know the consolidated consumption from its customers to compute the expected consumption and to keep the distribution network in equilibrium. Based on the consumptions of previous rounds, each energy supplier will estimate the consolidated consumption of the next round. Everyone expects that the virtualized power network will run without a power outage. Thus, the suppliers should find an estimated consumption sufficiently close to their customers' real consumption to keep the power network running without solutions based on curtailment, which waste energy. Figure 4.2 depicts multiple energy suppliers virtualizing a power network and their necessity of frequent consolidated consumption to compute better-expected consumption. The red represents an energy supplier with its customers and their expected consumption, its amount of generated electricity, and their real consumption, the green represents the others. Since the electricity price changes constantly, the suppliers aim for more than balancing the power network. They aim to generate for the real consumption. Thus, the expected consumption should be equal to the real consumption and the generation should be slightly bigger, but only enough to cover the transmission cost. If a supplier expects that their customers will have more consumption than other customers and the electricity generation is equal for both suppliers, but at the balance, the real consumption of their customers is inverted in relation to the expected, then the suppliers will have a dispute and probably a litigation. To solve this problem, they need to receive the consolidated consumptions with higher frequency to compute better estimations. The more frequent the measurements, the better the expected consumption.

There are many advantages for distributed small power sources [4], but without virtualization, the unique supplier can work as a broker for household generators. At the end, such a supplier is the unique buyer for the electricity generators and the unique seller for customers that characterizes a monopoly.

4.1.3 Fair Distribution

In a non-smart grid, the supplier can determine areas to supply with a commodity and areas without supply. For example, an energy supplier can determine areas with electricity and areas without it. In a smart grid, the supplier has more information and can determine the minimum and maximum consumption per customer. Moreover, the supplier can have different prices based on the amount of the consumption. Therefore, a smart grid can give us a quantity-based pricing with time-based pricing. They are dynamic in two dimensions, i.e., the price floats over time and over the amount consumed.

For example, the energy supplier can ensure the minimum amount of electricity for every customer. Its price might be very low. This threshold can change over time due to the generation. The supplier might set many different prices and even define a continuous price function. Figure 4.3 depicts an example of quantity-based pricing with four different prices, namely, minimum price, price 1, price 2, and maximum for minimal consumption range, above and below the average, and maximal consumption range, respectively.

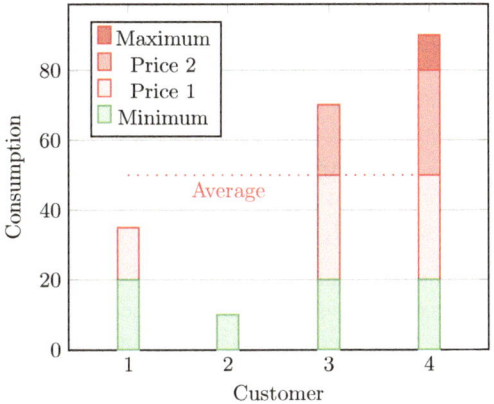

Fig. 4.3 Example of quantity-based pricing

The energy supplier can also identify consumption approaching a technical maximum and request the reduction of consumption before the protective relays are switched. Similarly, the water supplier can avoid low pressure or even prevent air from entering the pipes. In this scenario, the supplier could know the number of meters per range for the supplier to control whether they are reducing consumption. However, the supplier needs to know the consolidated consumption per range to balance the price per range over the time. Therefore, the supplier can control whether meters are approaching the maximum. In every PPP, supplier and their customers should be able to verify whether the values are correct. However, fair distribution might be an opportunity or misfortune depending on the law. The customer 4 could buy quotas from the customer 2 in Fig. 4.3. If it is legal, this is a new market. If not, and the meters are untrusted machines, then the transaction is not detected by comparison between the aggregated measurement a_j with consolidated consumptions c_j per range. Thus, it might blur the detection of consumption approaching a technical maximum. In contrast, if the meters are trusted, the legislation can fix quotas per customer. Therefore, independent of PPP, this application requires trusted meters to work accurately.

4.2 Requirements

This section presents the requirements for PPPs in a smart grid scenario. Besides security and privacy, PPPs have four minimum requirements [1].

Recoverability of Bills $b_i^\$$, i.e., the possibility of invoice recovery. The billing with
    time-based pricing is a requirement for smart grids [2], cf. Sect. 1.1. The billing
    with predetermined pricing is already required for a non-smart grid.
Recoverability of Consolidated consumptions c_j, i.e., the possibility of retrieving
    consolidated consumptions. The total consumption of all customers in a round
    is required for detection of overload [5], detections of fraud and energy
    loss, virtualization of the supplier commodity network, and fair distribution,
    cf. Sects. 4.1.1–4.1.3.
Verification (auditability), i.e., suppliers and their customers want to verify whether
    the computations are done correctly. Verification should be done to avoid fraud
    and errors. Verification is also known as non-interactive zero-knowledge proof.
    It is only effective with digital signatures to ensure non-repudiation.
Computational efficiency, i.e., the PPPs should demand few computational re-
    sources, both in processing time and in communication. Note that a PPP might
    be efficient for a specific number of meters but might be not scalable. The
    concept of efficiency depends on several variables.

The bill is already known for a non-smart grid, the old supplier model. However, the commodities can be very cheap in a short interval. Thus, if the supplier can apply time-based pricing, customers can benefit from low prices and the supplier sells more. The consolidated consumption may be used for fraud and loss detection.

In addition, it can be used for the virtualization of the supplier commodity network. Fair distribution needs more than consolidated consumptions c_j, but it can also be computed from the encrypted measurements $\mathfrak{M}_{i,j}$ and preserve all properties from the PPPs, if the meters are trusted. Verification is necessary to ensure safety and security. It can be used to avoid disputes between suppliers in a virtualized commodity network, or between supplier and its customers regarding the bill and billed consumption, or even to detect problems in the consolidated consumption. Note that verifications with digital signatures imply non-repudiation.

Moreover, the correct values are necessary and sufficient for the equations that govern the PPPs to hold. It uses the abbreviation **iff** "if and only if" representing a condition necessary and sufficient. Thus, the statement the equations hold **iff** the values are correct is equivalent to the values are correct **iff** the equations hold. Therefore, the equations hold if the values are correct and the values are correct if the equations hold. Verification as well as security and privacy should be ensured mathematically, i.e., an attacker should solve an intractable mathematical problem to change values or to get information from the system. Whereas the literature has no result whether a one-way function exists, the properties of the protocols are ensured by the assumption that a mathematical problem is intractable or infeasible to solve. Apart from the intractability, all computations in PPPs should be efficient.

Table 4.1 shows the private measurements in red at its center. The meters i and rounds j are in the first column and line, respectively, while the monetary value of the consolidated consumption c_j and the billed consumption b_i are in the last line and column, respectively. They are shaded to highlight the only part that the supplier needs to know. The sum of the monetary value of consolidated consumptions c_j should be equal to the sum of the billed consumption b_i, i.e.,

$$\sum_{i=1}^{\tilde{\imath}} b_i^{\$} = \sum_{j=1}^{\tilde{\jmath}} \text{Value}\left(c_j\right) = \sum_{j=1}^{\tilde{\jmath}} \text{Value}\left(a_j - \epsilon_j\right) = \sum_{i=1}^{\tilde{\imath}}\sum_{j=1}^{\tilde{\jmath}} \text{Value}\left(m_{ij}\right),$$

where the function Value determines the monetary value. Considering only the consumption in watts, we have

$$\sum_{i=1}^{\tilde{\imath}} b_i = \sum_{j=1}^{\tilde{\jmath}} c_j = \sum_{j=1}^{\tilde{\jmath}} a_j - \epsilon_j = \sum_{i=1}^{\tilde{\imath}}\sum_{j=1}^{\tilde{\jmath}} m_{i,j}. \tag{4.1}$$

Equation (4.1) holds for correct billed consumptions b_i and consolidated consumptions c_j. If a meter sent incorrect measurements, its supplier can determine by means of Eq. (4.1) that some values are incorrect or there is energy loss. However, the supplier cannot determine with Eq. (4.1) which meter sent incorrect values. Nevertheless, if a meter sent huge measurements causing disruption in the communication, the supplier can detect the faulty meter in the billing process, because the sum of its measurements will also be huge. From Table 4.1, the supplier can identify the meters and in which rounds they sent huge values.

Table 4.1 Consolidated consumption c_j versus billed consumption b_i

Meter \ Round	1	2	$\cdots$	j	$\cdots$	$\tilde{j}$	b_i
Meter 1	$m_{1,1}$	$m_{1,2}$	$\cdots$	$m_{1,j}$	$\cdots$	$m_{1,\tilde{j}}$	$\sum_{j=1}^{\tilde{j}} m_{1,j}$
Meter 2	$m_{2,1}$	$m_{2,2}$	$\cdots$	$m_{2,j}$	$\cdots$	$m_{2,\tilde{j}}$	$\sum_{j=1}^{\tilde{j}} m_{2,j}$
$\vdots$	$\vdots$	$\vdots$	$\ddots$	$\vdots$	$\ddots$	$\vdots$	$\vdots$
Meter i	$m_{i,1}$	$m_{i,2}$	$\cdots$	$m_{i,j}$	$\cdots$	$m_{i,\tilde{j}}$	$\sum_{j=1}^{\tilde{j}} m_{i,j}$
$\vdots$	$\vdots$	$\vdots$	$\ddots$	$\vdots$	$\ddots$	$\vdots$	$\vdots$
Meter $\tilde{\imath}$	$m_{\tilde{\imath},1}$	$m_{\tilde{\imath},2}$	$\cdots$	$m_{\tilde{\imath},j}$	$\cdots$	$m_{\tilde{\imath},\tilde{j}}$	$\sum_{j=1}^{\tilde{j}} m_{\tilde{\imath}\,j}$
c_j	$\sum_{i=1}^{\tilde{\imath}} m_{i,1}$	$\sum_{i=1}^{\tilde{\imath}} m_{i,2}$	$\cdots$	$\sum_{i=1}^{\tilde{\imath}} m_{i,j}$	$\cdots$	$\sum_{i=1}^{\tilde{\imath}} m_{i,\tilde{j}}$	$\sum_{j=1}^{\tilde{j}} c_j = \sum_{i=1}^{\tilde{\imath}} b_i$

PPPs aggregate the measurements but do not determine a secure value to $\tilde{\imath}$ and $\tilde{j}$. Note that an attacker trying to recover the measurements from meters knowing only billed consumption b_i has the same difficulty of recovering the measurements from a round knowing only consolidated consumption c_j. However, the difficulty is smaller when the attacker knows both b_i and c_j. The next chapter presents limitations of aggregation approach in two sections exploring algebraic properties and exploiting probabilistic properties.

References

1. F. Borges et al., A privacy-enhancing protocol that provides innetwork data aggregation and verifiable smart meter billing, in *2014 IEEE Symposium on Computers and Communication (ISCC)* (2014), pp. 1–6. doi:10.1109/ISCC.2014.6912612
2. P. Fox-Penner, *Smart Power: Climate Change, the Smart Grid, and the Future of Electric Utilities* (Island Press, Washington, DC, 2010). isbn:9781597268097
3. M. Jawurek, M. Johns, F. Kerschbaum, Plug-in privacy for smart metering billing, in *Proceedings of Privacy Enhancing Technologies: 11th International Symposium, PETS 2011, Waterloo, ON, July 27–29, 2011*, ed. by S. Fischer-Hübner, N. Hopper (Springer, Berlin, Heidelberg, 2011), pp. 192–210. isbn:978-3-642-22263-4. doi:10.1007/978-3-642-22263-4_11. http://dx.doi.org/10.1007/978-3-642-22263-4_11
4. A.B. Lovins et al., *Small is Profitable. The Hidden Economic Benefits of Making Electrical Resources the Right Size* (Rocky Mountain Institute, Boulder, CO, 2002). isbn:1-881071-07-3
5. K.D. McBee, M.G. Simoes, General smart meter guidelines to accurately assess the aging of distribution transformers. IEEE Trans. Smart Grid **5**(6), 2967–2979 (2014). issn:1949-3053. doi:10.1109/TSG.2014.2320285

6. M. Nicolosi, Energy efficiency policies and strategies with regular papers. Energy Policy **38**(11), 7257–7268 (2010). issn:0301-4215. doi:10.1016/j.enpol.2010.08.002. http://www.sciencedirect.com/science/article/pii/S0301421510005860

7. S. Salinas, M. Li, P. Li, Privacy-preserving energy theft detection in smart grids: a P2P computing approach. IEEE J. Sel. Areas Commun. **31**(9), 257–267 (2013). issn:0733-8716. doi:10.1109/JSAC.2013.SUP.0513023

Chapter 5
Quantifying the Aggregation Size

Abstract This chapter analyzes the possibility for an attacker to recover either individual measurements or probable individual measurements after the aggregations with any Privacy-Preserving Protocol (PPP). The relation between the measurements and the leak of privacy depends on several variables.

Keywords Leakage • Error-correcting code • Combination • Binomial • System of linear equations • Probability • Probable solutions

Savi et al. [7] presented an analysis on schemes based on noise to quantify a trade-off between the number of measurements that compound the consolidated consumption c_j and the precision on c_j. Some previous work used differential privacy [2] to analyze a specific PPP, for instance, the work of Jawurek and Kerschbaum [5]. Bohli et al. [1] presented a model for measuring the degree of privacy by means of a cryptographic game. However, the game does not consider all variables involved, cf. Sect. 3.1.3.

This chapter presents an analysis independent of PPP to quantify the leakage of information, which depends on the interval between the rounds j [3, 4, 6], on aggregation size—i.e., the number of smart meters $\tilde{\imath}$, cf. Sect. 2.2.3, on the number of rounds $\tilde{\jmath}$, cf. Sect. 4.2, and on the bit-length of consolidated consumptions c_j and billed consumptions b_i. The analysis is valid for all PPP that has two aggregations, i.e., it behaves like in Table 4.1 providing billed consumption b_i and consolidated consumption c_j. This analysis results in two kinds of properties, namely algebraic and probabilistic. The former can be used as an error-correcting code for the supplier. The latter shows how to approach a valid set of measurements and how to find all possible solutions. The difficulty for the attacker is to identify the correct solution. Nevertheless, many possible solutions can be excluded due to the timestamp of the rounds, the consumption pattern from a set of customers in previous days, weeks, etc.

© Springer International Publishing Switzerland 2017   49
F. Borges de Oliveira, *On Privacy-Preserving Protocols for Smart Metering Systems*,
DOI 10.1007/978-3-319-40718-0_5

5.1 Algebraic Properties

Assume that the attacker has three measurements per billed consumption and the unit of electricity is given by star ($\star$). Thus, the attacker tries to split up the consolidated consumption c_j into three boxes. If $c_j = 6$, one possible solution is $\boxed{\star\;\star\;\star}\;\boxed{\star}\;\boxed{\star\;\star}$. To simplify the formulation, instead of box, the stars can be split by bars. Thus, $\star\;\star\;\star\;|\;\star\;|\;\star\;\star$ has the same solution. With the star bar notation, the possible number of solutions is determined by the combination of 6 stars plus 2 bar choose 6 stars, which is given by

$$\binom{6+2}{6} = \frac{8!}{6!(8-6)!} = 28.$$

In general, for $\tilde{j}$ rounds and an arbitrary billed consumption b_i, the number of solutions for the attacker is determined by

$$\binom{b_i + \tilde{j} - 1}{\tilde{j}} = \frac{(b_i + \tilde{j} - 1)!}{(b_i - 1)!\,\tilde{j}\,!} = \binom{b_i + \tilde{j} - 1}{b_i - 1}. \tag{5.1}$$

Similarly, if the attacker has only the number of meters $\tilde{i}$ and the consolidated consumption c_j, the number of solutions is given by

$$\binom{c_j + \tilde{i} - 1}{\tilde{i}} = \frac{(c_j + \tilde{i} - 1)!}{(c_j - 1)!\,\tilde{i}\,!} = \binom{c_j + \tilde{i} - 1}{c_j - 1}. \tag{5.2}$$

The binomial of Eqs. (5.1) and (5.2) are known in textbooks as multichoose, multiset number, composition, and stars and bars.

The behavior of Eq. (5.1) is similar to Eq. (5.2). The former says that the attacker should try a number of possibilities in function of the total number of rounds $\tilde{j}$ and the sum given by billed consumption b_i. Similarly, the latter says that the number of possibilities is the function of the total number of meters $\tilde{i}$ and the sum given by consolidated consumption c_j. Figure 5.1 depicts the number of possibilities in relation to the sum, i.e., b_i or c_j, and the total, i.e., $\tilde{i}$ or $\tilde{j}$ with the values going from 0 to 10. Figure 5.1b depicts the contour plot of Fig. 5.1a. Figure 5.2 depicts the evolution of the possibilities growing with the values going from 0 to 20. Figure 5.3 depicts the ultimate behavior of Eqs. (5.1) and (5.2) with the logarithm of the possibilities to base 2 as a function of the sum (b_i or c_j) and the total ($\tilde{i}$ or $\tilde{j}$) computed with values from 1 to 2500. Comparing Figs. 5.1, 5.2, 5.3 with each other, we can see that their asymptotic growth only explodes in the last values of their domains. Comparing Figs. 5.1b, 5.2b, 5.3b with each other, we can see that the safe high probabilities are in a narrow range. Moreover, it becomes narrower when the sum (b_i or c_j) and the total ($\tilde{i}$ or $\tilde{j}$) increase. Therefore, the sum (b_i or c_j) and the total ($\tilde{i}$ or $\tilde{j}$) should be as close as possible to maximize the security.

Stirling's formula gives us an approximation for factorials, i.e., for an integer number n, we have

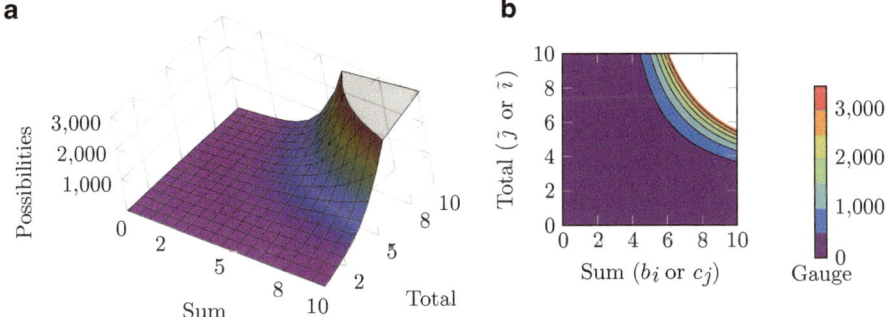

Fig. 5.1 Possibilities in relation to the sum (b_i or c_j) and the total ($\tilde{i}$ or $\tilde{j}$) up to 10. (**a**) Curve of possibilities. (**b**) Contour plot

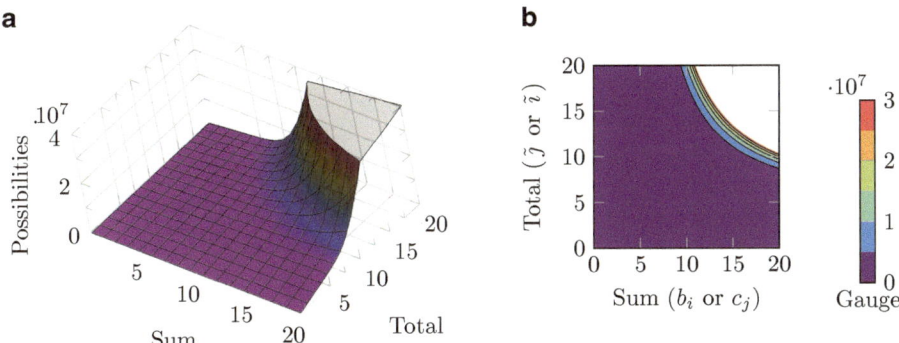

Fig. 5.2 Possibilities in relation to the sum (b_i or c_j) and the total ($\tilde{i}$ or $\tilde{j}$) up to 20. (**a**) Curve of possibilities. (**b**) Contour plot

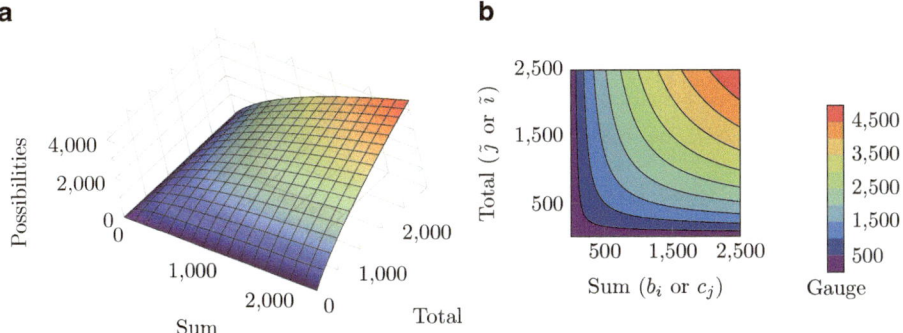

Fig. 5.3 Possibilities in relation to the sum (b_i or c_j) and the total ($\tilde{i}$ or $\tilde{j}$) in logarithmic scale. (**a**) Curve of possibilities. (**b**) Contour plot

$$\ln(n!) = \ln(1) + \ln(2) + \cdots + \ln(n)$$

$$= \sum_{k=1}^{n} \ln(k)$$

$$\approx \int_{1}^{n} \ln(x)\, dx. \qquad \text{integrating by parts} \qquad (5.3)$$

$$= [x\ln(x) - x]_{1}^{n}$$

$$= n\ln(n) - n + 1$$

$$\approx n\ln(n) - n.$$

Thus, we can apply his formula—Eq. (5.3)—to understand Eqs. (5.1) and (5.2) better. Since they have the same behavior, we can just choose one. Without loss of generality, let us calculate a logarithm in Eq. (5.1). The result is given by

$$\ln\binom{b_i + \tilde{j} - 1}{\tilde{j}} = \ln\frac{(b_i + \tilde{j} - 1)!}{(b_i - 1)!\,\tilde{j}\,!} \qquad (5.4)$$

$$= \ln((b_i + \tilde{j} - 1)!) - \ln((b_i - 1)!) - \ln(\tilde{j}\,!).$$

Using Stirling's formula in the result of Eq. (5.4), we find

$$\ln\binom{b_i + \tilde{j} - 1}{\tilde{j}} = \ln((b_i + \tilde{j} - 1)!) - \ln((b_i - 1)!) - \ln(\tilde{j}\,!)$$

$$= (b_i + \tilde{j} - 1)\ln(b_i + \tilde{j} - 1) + (b_i + \tilde{j} - 1)$$

$$- (b_i - 1)\ln(b_i - 1) - (b_i - 1) - \tilde{j}\,\ln(\tilde{j}\,) - \tilde{j} \qquad (5.5)$$

$$= (b_i + \tilde{j} - 1)\ln(b_i + \tilde{j} - 1)$$

$$- (b_i - 1)\ln(b_i - 1) - \tilde{j}\,\ln(\tilde{j}\,).$$

Equation (5.5) can provide an approximation for the bit-length of the number of combinations. Assume that $b_i - 1$ is approx. half of $\tilde{j}$, then Eq. (5.5) results

$$1.5\,\tilde{j}\,\ln(1.5\,\tilde{j}\,) - 0.5\,\tilde{j}\,\ln(0.5\,\tilde{j}\,) - \tilde{j}\,\ln(\tilde{j}\,)$$

$$= 1.5\,\tilde{j}\,\ln(1.5\,\tilde{j}\,) - \ln((0.5\,\tilde{j}\,)^{0.5\,\tilde{j}}\,\tilde{j}^{\,\tilde{j}})$$

$$= \ln\left(\frac{1.5^{1.5\,\tilde{j}}\,\tilde{j}^{\,1.5\,\tilde{j}}}{0.5^{0.5\,\tilde{j}}\,\tilde{j}^{\,1.5\,\tilde{j}}}\right) \qquad (5.6)$$

$$= \ln\left(\frac{1.5^{1.5\,\tilde{j}}}{0.5^{0.5\,\tilde{j}}}\right) \approx \ln\left(\left(\frac{1.8}{0.7}\right)^{\tilde{j}}\right) \approx \ln(2.6^{\tilde{j}})$$

$$= \ln(2^{\log_2(2.6)\,\tilde{j}}) \approx \ln(2^{1.4\,\tilde{j}}).$$

Differently, assume that $b_i - 1$ is approx. $\tilde{j}$, then Eq. (5.5) results

$$2\tilde{j} \ln(2\tilde{j}) - \tilde{j} \ln(\tilde{j}) - \tilde{j} \ln(\tilde{j})$$
$$= 2\tilde{j} \ln(2\tilde{j}) - 2\tilde{j} \ln(\tilde{j})$$
$$= \ln\left(\left(\frac{2\tilde{j}}{\tilde{j}}\right)^{2\tilde{j}}\right) \tag{5.7}$$
$$= \ln(2^{2\tilde{j}}) = \ln(4^{\tilde{j}}).$$

Contrarily, assume that $b_i - 1$ is approx. double $\tilde{j}$, then Eq. (5.5) results

$$3\tilde{j} \ln(3\tilde{j}) - 2\tilde{j} \ln(2\tilde{j}) - \tilde{j} \ln(\tilde{j})$$
$$= 3\tilde{j} \ln(3\tilde{j}) - \ln((2\tilde{j})^{2\tilde{j}} \tilde{j}^{\tilde{j}})$$
$$= \ln\left(\frac{3^{3\tilde{j}} \tilde{j}^{3\tilde{j}}}{2^{2\tilde{j}} \tilde{j}^{3\tilde{j}}}\right) \tag{5.8}$$
$$= \ln\left(\frac{3^{3\tilde{j}}}{2^{2\tilde{j}}}\right) = \ln\left(\left(\frac{27}{4}\right)^{\tilde{j}}\right) = \ln(6.75^{\tilde{j}})$$
$$= \ln(2^{\log_2(6.75)\tilde{j}}) \approx \ln(2^{2.8\tilde{j}}).$$

If we divide Eqs. (5.6)–(5.8) by $\ln(2)$, we have $1.4\tilde{j}$, $2\tilde{j}$, and $2.8\tilde{j}$ bits, respectively. From half to double, we increased $b_i - 1$ by 4 times to get the double number of bits. In a limited interval to $b_i - 1$ and j, the number of bits tends to maximum when $b_i - 1$ tends j. Similarly, when we calculate $\tilde{j}$ or $b_i - 1$ tending to zero, the number of bits tends to zero. Since Eq. (5.1) is syntactically equal to Eq. (5.2), the same results are valid for c_j and $\tilde{i}$. These theoretical results are in agreement with the experimental results presented in Fig. 5.3.

In summary, Figs. 5.1 and 5.2 depict that the maximum is achieved with $\tilde{i} = c_j$ and $\tilde{j} = b_i$, respectively. However, not all possibilities are solutions of the system of linear equations. Figure 5.3 depicts the curve with respect to the number of bits. Figures 5.1 and 5.2 depict the growing number of possibilities with a narrow range like a rainbow where the number of possibilities is bigger, therefore, more interesting. Figure 5.3b depicts this narrow range with respect to the number of bits.

The number of combinations necessary for an attacker to discover all measurements used to compute the billed consumption b_i or the consolidated consumption c_j is given by Eq. (5.1), if b_i and $\tilde{j}$ are known, and by Eq. (5.2), if consolidated consumption c_j and number of users $\tilde{i}$ are known. However, if these values are known, the attacker can speed up the search for the individual measurements. Firstly, let $\tilde{i} = \tilde{j} = 2$. Thus, Table 4.1 gives us the system of linear equations

$$\begin{cases} b_1 = m_{1,1} + m_{1,2} \\ b_2 = m_{2,1} + m_{2,2} \\ c_1 = m_{1,1} + m_{2,1} \\ c_2 = m_{1,2} + m_{2,2}. \end{cases}$$

These equations are linearly dependent, namely $c_2 = b_1 + b_2 - c_1$. Thus, we can eliminate the last equation from the system and write it in a matrix form, i.e.,

$$\begin{vmatrix} b_1 \\ b_2 \\ c_1 \end{vmatrix} = \begin{vmatrix} 1 & 1 & 0 & 0 \\ 0 & 0 & 1 & 1 \\ 1 & 0 & 1 & 0 \end{vmatrix} \begin{vmatrix} m_{1,1} \\ m_{1,2} \\ m_{2,1} \\ m_{2,2} \end{vmatrix}. \tag{5.9}$$

The system has 3 equations and 4 unknowns. Hence, it has an infinite number of solutions for the set of the real numbers $\mathbb{R}$. Nevertheless, the number of solutions for the set of the natural numbers $\mathbb{N}$ is finite. Moreover, the system has a unique solution if one measurement is known. Equation (5.9) shows that if $m_{2,2}$ is known, the number of equations is equal to the number of unknowns, and therefore, the system has a unique solution. One can compute different linear combinations to obtain the solution of the system, if another measurement is known. An important question is raised and we wonder how many measurements an attacker needs to know to solve bigger systems. Before the general case, $\tilde{\imath} = \tilde{\jmath} = 3$. Thus, the new system of linear equation in matrix notation with dots to simplify the visualization is given by

$$\begin{vmatrix} b_1 \\ b_2 \\ b_3 \\ c_1 \\ c_2 \\ c_3 \end{vmatrix} = \begin{vmatrix} 1 & 1 & 1 & 0 & 0 & 0 & 0 & 0 & 0 \\ 0 & 0 & 0 & 1 & 1 & 1 & 0 & 0 & 0 \\ 0 & 0 & 0 & 0 & 0 & 0 & 1 & 1 & 1 \\ 1 & 0 & 0 & 1 & 0 & 0 & 1 & 0 & 0 \\ 0 & 1 & 0 & 0 & 1 & 0 & 0 & 1 & 0 \\ 0 & 0 & 1 & 0 & 0 & 1 & 0 & 0 & 1 \end{vmatrix} \begin{vmatrix} m_{1,1} \\ m_{1,2} \\ m_{1,3} \\ m_{2,1} \\ m_{2,2} \\ m_{2,3} \\ m_{3,1} \\ m_{3,2} \\ m_{3,3} \end{vmatrix}. \tag{5.10}$$

Equation (5.10) shows that $c_3 = b_1 + b_2 + b_3 - c_1 - c_2$. Thus, the last line of the matrix of known values and of the binary matrix can be eliminated. We cannot eliminate more lines because the rank of the binary matrix is 5, i.e., there is not more dependence. Consequently, an attacker needs to know 4 measurements to solve the system. However, the knowledge of 3 measurements from the same smart meter reduces the rank of the binary matrix. In contrast, the knowledge of the measurements that compose c_3 does not change the matrix rank. The attacker needs to choose one more measurement to solve the system algebraically.

In general, the system of equations is given by

$$
\begin{vmatrix} b_1 \\ b_2 \\ \vdots \\ b_{\tilde{\imath}} \\ c_1 \\ c_2 \\ \vdots \\ c_{\tilde{\jmath}} \end{vmatrix} = \begin{vmatrix} 1\,1\,\cdots\,1 & 0\,0\,\cdots\,0 & \cdots & 0\,0\,\cdots\,0 \\ 0\,0\,\cdots\,0 & 1\,1\,\cdots\,1 & \cdots & 0\,0\,\cdots\,0 \\ \vdots\,\vdots\,\ddots\,\vdots & \vdots\,\vdots\,\ddots\,\vdots & \ddots & \vdots\,\vdots\,\ddots\,\vdots \\ 0\,0\,\cdots\,0 & 0\,0\,\cdots\,0 & \cdots & 1\,1\,\cdots\,1 \\ 1\,0\,\cdots\,0 & 1\,0\,\cdots\,0 & \cdots & 1\,0\,\cdots\,0 \\ 0\,1\,\cdots\,0 & 0\,1\,\cdots\,0 & \cdots & 0\,1\,\cdots\,0 \\ \vdots\,\vdots\,\ddots\,\vdots & \vdots\,\vdots\,\ddots\,\vdots & \ddots & \vdots\,\vdots\,\ddots\,\vdots \\ 0\,0\,\cdots\,1 & 0\,0\,\cdots\,1 & \cdots & 0\,0\,\cdots\,1 \end{vmatrix} \begin{vmatrix} m_{1,1} \\ m_{1,2} \\ \vdots \\ m_{1,\tilde{\jmath}} \\ m_{2,1} \\ m_{2,2} \\ \vdots \\ m_{\tilde{\imath},\tilde{\jmath}} \end{vmatrix}.
$$
(5.11)

One can see that

$$
c_{\tilde{\jmath}} = \sum_{i=1}^{\tilde{\imath}} b_i - \sum_{i=1}^{\tilde{\jmath}-1} c_j.
$$

In addition, the matrix rank is $\tilde{\imath} + \tilde{\jmath} - 1$ and the number of unknowns is $\tilde{\imath} \cdot \tilde{\jmath}$. Therefore, an attacker should know η measurements s.t.

$$
\eta = \tilde{\imath} \cdot \tilde{\jmath} - \tilde{\imath} - \tilde{\jmath} + 1
$$
(5.12)

to solve the system. Nevertheless, some measurements might reduce the rank. Equation (5.12) shows that the difficulty for solving Eq. (5.11) grows with $\tilde{\imath}$ and $\tilde{\jmath}$. In particular, the maximum number of possibilities is achieved with $\tilde{\imath} = \tilde{\jmath}$. Figure 5.4 depicts the difficulty growing presented by Eq. (5.12).

For an example with $\tilde{\imath} > \tilde{\jmath}$, consider

$$
\begin{vmatrix} b_1 \\ b_2 \\ b_3 \\ c_1 \\ c_2 \end{vmatrix} = \begin{vmatrix} 1\,1\,0\,0\,0\,0 \\ 0\,0\,1\,1\,0\,0 \\ 0\,0\,0\,0\,1\,1 \\ 1\,0\,1\,0\,1\,0 \\ 0\,1\,0\,1\,0\,1 \end{vmatrix} \begin{vmatrix} m_{1,1} \\ m_{1,2} \\ m_{2,1} \\ m_{2,2} \\ m_{3,1} \\ m_{3,2} \end{vmatrix}.
$$
(5.13)

Contrarily, for an example with $\tilde{\imath} < \tilde{\jmath}$, consider

$$
\begin{vmatrix} b_1 \\ b_2 \\ c_1 \\ c_2 \\ c_3 \end{vmatrix} = \begin{vmatrix} 1\,1\,1\,0\,0\,0 \\ 0\,0\,0\,1\,1\,1 \\ 1\,0\,0\,1\,0\,0 \\ 0\,1\,0\,0\,1\,0 \\ 0\,0\,1\,0\,0\,1 \end{vmatrix} \begin{vmatrix} m_{1,1} \\ m_{1,2} \\ m_{1,3} \\ m_{2,1} \\ m_{2,2} \\ m_{2,3} \end{vmatrix}.
$$
(5.14)

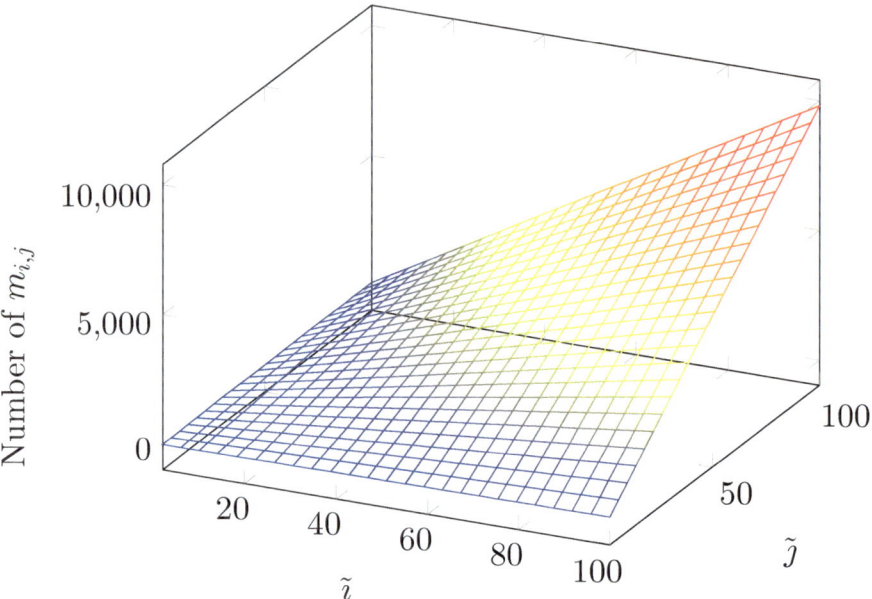

Fig. 5.4 Number of measurements $m_{i,j}$ necessary to solve the system

Note that matrices from Eqs. (5.13) and (5.14) have the same dimension, because the former has $\tilde{\imath} = 3$ and $\tilde{\jmath} = 2$, and the latter has $\tilde{\imath} = 2$ and $\tilde{\jmath} = 3$.

If the supplier loses either one billed consumption b_i or one consolidated consumption c_j, then it can recover the lost value. The system of equations works as an error-correcting code, i.e., it can detect and correct errors in the binary matrix and can recover either one b_i or one c_j without disclosing the measurements $m_{i,j}$.

5.2 Probabilistic Properties

The possible measurements $m_{i,j}$ may be found with probabilities in an easier and faster way than using algebra. One can create a function that returns the most probable value of the measurements $m_{i,j}$ based on the last line and column in Table 4.1, i.e., in the billed consumption b_i and the consolidated consumption c_j. Among other variables, the function might consider the timestamp to determine hours with less or more electricity consumption.

For simplicity, let us consider a simplified model based only on averages where the probable value of the measurement $m_{i,j}$ is given by

$$\overline{m_{ij_1}} \stackrel{\text{def.}}{=} \frac{b_i/\tilde{\jmath} + c_j/\tilde{\imath}}{2}. \tag{5.15}$$

Using Eq. (5.15), one can reconstruct Table 4.1. However, the sum of the measurements might give a different value from the billed consumption b_i and the consolidated consumption c_j indicating that the measurements found are not a solution for the system of equations. Nevertheless, the probable value $\overline{m_{ij}}$ can be used to find a better approximation with a new probable value until a probable value becomes equal to the measurement that satisfies the system of equations. For even values of k, the new approximation can be defined by

$$\overline{m_{ij_k}} \overset{\text{def.}}{=} \frac{\overline{m_{ij_{k-1}}}}{c_j/\overline{c_j}}, \tag{5.16}$$

where

$$\overline{c_{j_k}} \overset{\text{def.}}{=} \sum_{j=1}^{\tilde{j}} \overline{m_{ij_{k-1}}}.$$

For odd values of k, the new approximation can be defined by

$$\overline{m_{ij_k}} \overset{\text{def.}}{=} \frac{\overline{m_{ij_{k-1}}}}{b_i/\overline{b_i}}, \tag{5.17}$$

where

$$\overline{b_{i_k}} \overset{\text{def.}}{=} \sum_{i=1}^{\tilde{i}} \overline{m_{ij_{k-1}}}.$$

Equations (5.16) and (5.17) can be used recursively to determine a better approximation until $m_{ij} \approx \overline{m_{ij_k}}$ for all i and all j. Let us see a numerical example that starts with Table 5.1, where an attacker only knows the shaded part. Note that Eq. (5.16) adjusts the last line, i.e., the consolidated consumptions, while Eq. (5.17) adjusts the last column, i.e., the billed consumptions.

Table 5.1 Example of hidden measurements

	Round 1	Round 2	Round 3	Round 4	b_i
Meter 1	$m_{1,1}$	$m_{1,2}$	$m_{1,3}$	$m_{1,4}$	10
Meter 2	$m_{2,1}$	$m_{2,2}$	$m_{2,3}$	$m_{2,4}$	212
Meter 3	$m_{3,1}$	$m_{3,2}$	$m_{3,3}$	$m_{3,4}$	1,106
c_j	601	10	503	214	1,328

Applying Eq. (5.15) to the values from Table 5.1 and writing the result in a matrix, we have

$$
\begin{vmatrix}
101.42 & 2.92 & 85.08 & 36.92 & 226.33 \\
126.67 & 28.17 & 110.33 & 62.17 & 327.33 \\
238.42 & 139.92 & 222.08 & 173.92 & 774.33 \\
466.50 & 171 & 417.50 & 273 & 1,328
\end{vmatrix}
$$

whose coefficients are a rough approximation of a solution. To improve the approximation, we can apply Eq. (5.15), which results in

$$
\begin{vmatrix}
130.66 & 0.17 & 102.51 & 28.94 & 262.27 \\
163.19 & 1.65 & 132.93 & 48.73 & 346.49 \\
307.16 & 8.18 & 267.56 & 136.33 & 719.23 \\
601 & 10 & 503 & 214 & 1,328
\end{vmatrix}
.
$$

After applying Eq. (5.16), the values from the consolidated consumptions c_j are correct. However, we still do not have a solution because the values from the billed consumptions b_i are incorrect. Thus, applying Eq. (5.17), we have

$$
\begin{vmatrix}
4.98 & 0.01 & 3.91 & 1.10 & 10 \\
99.84 & 1.01 & 81.33 & 29.82 & 212 \\
472.33 & 12.58 & 411.45 & 209.64 & 1,106 \\
577.16 & 13.60 & 496.69 & 240.56 & 1,328
\end{vmatrix}
$$

whose last column for b_i is correct, but the last line for c_j is incorrect again. Nevertheless, we are approaching a solution that satisfies Table 5.1. Thus, applying Eq. (5.16) again, we have

$$
\begin{vmatrix}
5.19 & 0.00 & 3.96 & 0.98 & 10.13 \\
103.97 & 0.74 & 82.37 & 26.52 & 213.60 \\
491.84 & 9.25 & 416.68 & 186.49 & 1,104.27 \\
601 & 10 & 503 & 214 & 1,328
\end{vmatrix}
,
$$

which is very close to the next iteration with Eq. (5.17) given by

$$
\begin{vmatrix}
5.12 & 0.00 & 3.91 & 0.97 & 10 \\
103.19 & 0.74 & 81.75 & 26.33 & 212 \\
492.61 & 9.27 & 417.33 & 186.79 & 1,106 \\
600.93 & 10.01 & 502.99 & 214.08 & 1,328
\end{vmatrix}
.
$$

After five steps the matrix has converged, i.e., the next step gives an equivalent result. Therefore, the rounding of the coefficients gives us

$$\begin{vmatrix} 5 & 0 & 4 & 1 & 10 \\ 103 & 1 & 82 & 26 & 212 \\ 493 & 9 & 417 & 187 & 1,106 \\ 601 & 10 & 503 & 214 & 1,328 \end{vmatrix}.$$

After one finds a solution, it is easy to find the others by computing operations that preserve the sums, e.g.,

$$\begin{vmatrix} 5-3 & 0 & 4+3 & 1 & 10 \\ 103+3 & 1 & 82-3 & 26 & 212 \\ 493 & 9 & 417 & 187 & 1,106 \\ 601 & 10 & 503 & 214 & 1,328 \end{vmatrix}.$$

An attacker needs to use extra information to determine which solution is correct. Depending on the level of accuracy, an attacker can join consecutive consolidated consumptions in the same column and even split them to infer the correct solution.

For the line where $b_1 = 10$, the number of combinations can be determined by Eq. (5.1), and it is given by

$$\binom{10+4-1}{4} = 750.$$

Similarly, for the column where $c_2 = 10$, the number of combinations can be determined by Eq. (5.2), which results

$$\binom{10+3-1}{3} = 220.$$

However, not all combinations fit together and many of them can be excluded due to the consumption pattern. In addition, algebraic results can be used with probabilistic results to improve the attack. Moreover, a probable solution can always be quickly found. Therefore, the difficulty for the attacker is to recognize the most probable solutions.

References

1. J.-M. Bohli, C. Sorge, O. Ugus, A privacy model for smart metering, in *2010 IEEE International Conference on Communications Workshops (ICC)* (2010), pp. 1–5. doi:10.1109/ICCW.2010.5503916

2. C. Dwork, Differential privacy: a survey of results. English, in *Theory and Applications of Models of Computation*, ed. by M. Agrawal et al., vol. 4978. Lecture Notes in Computer Science (Springer, Berlin, Heidelberg, 2008), pp. 1–19. isbn:978-3-540-79227-7. doi:10.1007/978-3-540-79228-4_1. http://dx.doi.org/10.1007/978-3-540-79228-4_1
3. G. Eibl, D. Engel, Influence of data granularity on smart meter privacy. IEEE Trans. Smart Grid **6**(2), 930–939 (2015). issn:1949–3053. doi:10.1109/TSG.2014.2376613
4. U. Greveler, B. Justus, D. Löhr, Multimedia content identification through smart meter power usage profiles, in *Computers, Privacy and Data Protection (CPDP 2012)*. The Steering Committee of The World Congress in Computer Science, Computer Engineering and Applied Computing (WorldComp) (2012)
5. M. Jawurek, F. Kerschbaum, Fault-tolerant privacy- preserving statistics, in *Privacy Enhancing Technologies*, ed. by S. Fischer-Hübner, M. Wright, vol. 7384. Lecture Notes in Computer Science (Springer, Berlin, Heidelberg, 2012), pp. 221–238. isbn:978-3-642-31679-1. doi:10.1007/978-3-642-31680-7_12. http://dx.doi.org/10.1007/978-3-642-31680-7_12
6. A. Molina-Markham et al., Private memoirs of a smart meter, in *Proceedings of the 2nd ACM Workshop on Embedded Sensing Systems for Energy-Efficiency in Building*. BuildSys '10 (ACM, Zurich, 2010), pp. 61–66. isbn:978-1-4503-0458-0. doi:10.1145/1878431.1878446. http://doi.acm.org/10.1145/1878431.1878446
7. M. Savi, C. Rottondi, G. Verticale, Evaluation of the precision-privacy tradeoff of data perturbation for smart metering. IEEE Trans. Smart Grid **6**(5), 2409–2416 (2015). issn:1949-3053. doi:10.1109/TSG.2014.2387848

Chapter 6
Selected Privacy-Preserving Protocols

Abstract This chapter presents four Privacy-Preserving Protocols (PPPs)—PPP1 to PPP4—based on Symmetric DC-Nets (SDC-Nets), Elliptic Curve Cryptography (ECC), Asymmetric DC-Nets (ADC-Nets), and quantum cryptography, respectively. Besides efficiency, security, and privacy, the first protocol provides only the consolidated monetary value $c_j^\$$ while the second is designed only to provide billing based on dynamic pricing with verification of each bill $b_i^\$$. The third gives us the property of the two first protocols. Indeed, it provides all properties required in Sect. 4.2, namely: consolidated consumption, billing based on dynamic pricing, verification of aggregation and billing, and computational efficiency. Although the last protocol only provides the consolidated consumption, it pioneers PPPs based on quantum mechanics, i.e., this work presents the first PPP based on quantum mechanics to smart grids. In addition, quantum cryptography is more promising than quantum computers, and today, we already can buy devices that provide quantum cryptography.

Keywords Privacy-preserving protocols • Symmetric DC-Nets (SDC-Nets) • Elliptic curve cryptography • Asymmetric DC-nets (ADC-Nets) • Quantum cryptography • Verification • Efficiency

The four protocols presented in this chapter use a function to convert the measurements into monetary values. This function is important to simplify the protocols separating dynamic pricing from the security layer. Moreover, the security focus is to obtain the consolidated consumption in monetary value, i.e., consolidated monetary value. Therefore, the supplier can be abstracted as a counting agent and the protocols can be applied in other scenarios that require counting agents. Normally, the communication in the protocols for smart grid is described as Machine-to-Machine (M2M). Differently, customers with their smart meters are addressed as users in this chapter.

© Springer International Publishing Switzerland 2017   61
F. Borges de Oliveira, *On Privacy-Preserving Protocols for Smart Metering Systems*,
DOI 10.1007/978-3-319-40718-0_6

6.1 Monetary Value

The monetary value of a measurement $m_{i,j}$ is just the current price multiplied by the consumption. Nevertheless, the supplier has two prices: buying price $\underline{p}_j$ and selling price $\overline{p}_j$. Hence, users buy with the selling price $\overline{p}_j$ and sell with the buying price $\underline{p}_j$. The measurement $m_{i,j}$ might be measured in watts and can be positive for consumption and negative for generation. The signs can be inverted, but historically the consumption came first and it is given by a positive measurement. To transform a measurement to a monetary value, we can use the sign function. Note that $-(\text{sgn}(m_{i,j}) - 1)/2$ returns 0 or 1 when $m_{i,j}$ is positive or negative, respectively. In contrast, $(\text{sgn}(m_{i,j}) + 1)/2$ returns 1 or 0 when $m_{i,j}$ is positive or negative, respectively. We can use this observation to construct the function of monetary value given by

$$\text{Value}\left(m_{i,j}\right) \overset{\text{let}}{=} m_{i,j} \cdot \overline{p}_j^{-(\text{sgn}(m_{i,j})(m_{i,j})-1)/2} \, \underline{p}_j^{(\text{sgn}(m_{i,j})(m_{i,j})+1)/2}. \tag{6.1}$$

Equivalent to Eq. (6.1), we can write Algorithm 4, which behaves as the Value function. Note that in both cases, the $m_{i,j}$ is a multiple of the result. Thus, we do not need to address the case $m_{i,j} = 0$.

The monetary value is important to simplify the cryptographic algorithms. Certainly, the buying price $\underline{p}_j$ and the selling price $\overline{p}_j$ are arguments of the function in Eq. (6.35). However, they are omitted to keep a clean notation. In fact, the measurements are more important than these arguments in PPP descriptions. The buying price $\underline{p}_j$ and the selling price $\overline{p}_j$ are important to enable time-based pricing. However, the protocols can run normally without this feature. In other words, making the commodity price equal to 1 with $m_{i,j}$ always positive, we have $m_{i,j} = $ Value $\left(m_{i,j}\right)$, i.e., the protocols transmit the measurements without time-based pricing. The values of the buying price $\underline{p}_j$ and the selling price $\overline{p}_j$ should be public to avoid attacks. Otherwise, the counting agent could insert many zeros to get the value of one measurement $m_{i,j}$.

Algorithm 4: Monetary value

    Input: Measurement $m_{i,j}$, buying price $\underline{p}_j$, and selling price $\overline{p}_j$.

    Output: Monetary value of $m_{i,j}$.

1   **if** $m_{i,j} > 0$ **then**
2 | $v \leftarrow m_{i,j} \cdot \overline{p}_j$
3   **else**
4 | $v \leftarrow m_{i,j} \cdot \underline{p}_j$
5   **return** v

The function Value $(m_{i,j})$ can be more complex to satisfy the requirements of Sect. 4.1.3. Nevertheless, they can be implemented with IF instructions and concatenation of the values per range in a message.

6.2 PPP1 The Fastest

The fastest PPP presented in this book is PPP1, which is presented in [6] as part of a PPP to provide consolidated monetary value $c_j^\$$ without the Value function given by Eq. (6.1) nor Algorithm 4. PPP1 is based on SDC-Nets [13] with in-network aggregation, which generates a spanning tree to include all meters into the aggregation, i.e., the meters send encrypted measurements to each other until the last meter sends the encrypted consolidated consumption to their counting agent, as done in [5, 30, 31]. Without loss of generality, Fig. 6.1 presents only three users to depict the communication model using in-network aggregation. PPP1 assumes that the counting agent has an extra information channel to verify the aggregation. For example, energy suppliers can receive information from phasor measurement units (PMUs). In general, a supplier should know how much of a commodity it inserted in its supply network. In like manner, the number of voters in an election should be equal to the total number of votes cast for all candidates and the number of ballots. In the supplier scenarios, this information is called aggregated measurement a_j and is used to detect anomalies in the same round j before the supplier detects them in the billing process. Other PPPs could also consider aggregated measurements a_j.

With respect to the knowledge of the keys, SDC-Net can be represented with a graph structure. Instead of using a fully connected SDC-Net such as the protocols in Sect. 3.2.2 use, PPP1 uses a star SDC-Net with the counting agent in the center. Figure 6.2 depicts the key exchange between users and counting agent for a fully SDC-Net in Fig. 6.2a and for a star SDC-Net in Fig. 6.2b.

Because of shared keys, the counting agent can decrypt single measurements for the star SDC-Net and PPP1 works under the assumption that the users are honest-but-curious, and no attacker can spoof their communication in the aggregation process. Consequently, the encryption function is defined by

$$\mathfrak{M}_{i,j} \stackrel{\text{def.}}{=} \text{Enc}\left(m_{i,j}\right) \stackrel{\text{let}}{=} \text{Value}\left(m_{i,j}\right) + \text{H}\left(k_i\|j\right), \tag{6.2}$$

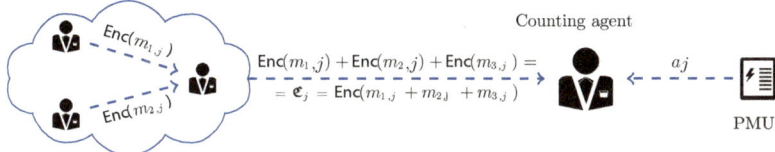

Fig. 6.1 In-network aggregation for three users

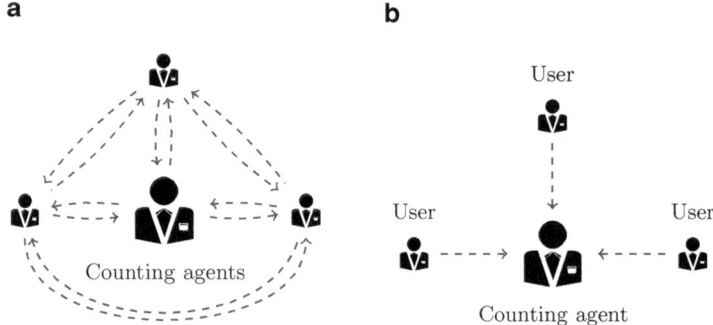

Fig. 6.2 Key exchange between users and their counting agent. (**a**) Fully connected SDC-Net. (**b**) Star SDC-Net

where $m_{i,j}$ is the measurement of the meter i in the round j, k_i is the key of the meter i known also by the counting agent, $||$ denotes string concatenation, and H is a secure hash function s.t. it behaves as a one-way function and has collision resistance. The aggregation is given by

$$\mathfrak{C}_j \overset{\text{def.}}{=} \sum_{i=1}^{\tilde{\imath}} \mathfrak{M}_{i,j}, \tag{6.3}$$

where $\mathfrak{M}_{i,j}$ is the encrypted measurement and $\tilde{\imath}$ is the number of users. Thus, the decryption function is given by

$$c_j^{\$} \overset{\text{def.}}{=} \mathsf{Dec}\left(\mathfrak{C}_j\right) \overset{\text{let}}{=} \mathfrak{C}_j - \sum_{i=1}^{\tilde{\imath}} \mathsf{H}\left(k_i || j\right), \tag{6.4}$$

where $\mathfrak{C}_j$ is the encrypted consolidated consumption computed using in-network aggregation.

Using the function Value, PPP1 runs independently of how prices float and Algorithm 5 can be written without caring about buying prices $\underline{p}_j$ and selling prices $\overline{p}_j$. The function can determine if a measurement represents consumption or generation by its sign $\text{sgn}(m_{i,j})$.

Algorithm 5 describes the three procedures of PPP1. Users calculate encryption and aggregation. Thus, each user runs an iteration of each loop. The counting agent who runs $\tilde{\imath}$ iterations calculates the decryption.

Algorithm 5: PPP1

---

1 **Procedure Encryption**

    Input: measurements $m_{i,j}$.

    Output: encrypted measurements $\mathfrak{M}_{i,j}$.

2   **for** $i \leftarrow 1$ **to** $\tilde{\imath}$ **do**

3          $\mathfrak{M}_{i,j} \leftarrow \text{Value}\left(m_{i,j}\right) + \mathsf{H}\left(k_i \| j\right)$

4 **Procedure Aggregation**

    Input: encrypted measurements $\mathfrak{M}_{i,j}$.

    Output: encrypted consolidated consumption $\mathfrak{C}_j$.

5      $\mathfrak{C}_j \leftarrow 0$

6   **for** $i \leftarrow 1$ **to** $\tilde{\imath}$ **do**

7          $\mathfrak{C}_j \leftarrow \mathfrak{C}_j + \mathfrak{M}_{i,j}$

8 **Procedure Decryption**

    Input: encrypted consolidated consumption $\mathfrak{C}_j$.

    Output: consolidated monetary value $c_j^{\$}$.

9      $c_j^{\$} \leftarrow \mathfrak{C}_j$

10   **for** $i \leftarrow 1$ **to** $\tilde{\imath}$ **do**

11          $c_j^{\$} \leftarrow c_j^{\$} - \mathsf{H}\left(k_i \| j\right)$

6.2.1 Security Analysis

The attacker model can be simplified by considering that the counting agent is the attacker whose goal is to get information about the measurements. The counting agent can use all information to recover measurements or keys but cannot collude with a user. A secure hash function s.t. it behaves as a one-way function and has collision resistance ensures that each user i has a pseudo-random number for every round j. Moreover, such numbers are used only once. PPP1 security relies on a secure hash function. Fast symmetric encryption algorithms can be used as a hash function H, i.e., an algorithm that behaves as a one-way function. The only one that can decrypt the measurements is the counting agent, and Algorithm 5 works similar to Algorithm 1, i.e., Eq. (6.2) works as Eq. (3.2) for encryption functions, Eq. (6.3) works as Eq. (3.3) for aggregations, and Eq. (6.4) works as Eq. (3.4) for decryption functions. Therefore, the star SDC-Net works as an additive homomorphic encryption primitive (AHEP).

6.2.2 Privacy Analysis

Maintaining privacy means keeping individual measurements inaccessible. Thus, no one should be able to recover k_i from $\mathsf{H}\left(k_i \| j\right)$ with H being a secure hash function s.t. it behaves as a one-way function and has collision resistance. Thus, users must not be able to eavesdrop on each other's measurements. The counting agent has the keys k_i, but only receives the encrypted consolidated consumption. As an AHEP, the privacy of PPP1 is ensured by in-network aggregation.

6.2.3 Performance Analysis

In-network aggregation enables users to send the minimum number of messages and the counting agent receives the minimum number of messages. Thus, the overhead in the communication network is optimal. However, the processing time for decryption grows with the number of users. Instead of the counting agent receives and processes one message from each user, it needs to process one hash function per user. Differently, AHEPs have their complexities based on cryptographic trapdoor functions, and their complexities do not depend on the number of users.

Using PPP1, users calculate almost only one hash function H, which is faster than any asymmetric cryptographic primitive is. Start SDC-Net has the minimum connectivity. Certainly, one can create a scheme in which not all messages will be encrypted. However, such a strategy requires more assumptions.

On the whole, PPP1 requires the computation of $2\tilde{\imath}$ hash functions while protocols based on fully connected SDC-Nets require the order of $\tilde{\imath}^2$ hash functions. Thus, PPP1 is quadratically faster than protocols using fully connected SDC-Net, e.g., LOP, cf. Sect. 3.2.2. Paillier is a fast AHEP and requires more than $2\tilde{\imath}$ modular exponentiations whose processing time depends on the exponent size [28]. The argument $k_i \| j$ is much smaller than n from Eq. (3.2). Indeed, the bit length of $k_i \| j$ can be less than half than the bit length of a scalar for ECC, which is well known for having small keys and, consequently, for being faster than other asymmetric primitives. As an expected result, PPP1 is faster than protocols based on well-known asymmetric primitives.

It is not possible to create an SDC-Net less connected than a fully connected SDC-Net without assuming that users trust each other. For this reason, PPP1 assumes the honest-but-curious trust model. Similarly, it is not possible to create an SDC-Net less connected than a star SDC-Net without assuming that an attacker cannot eavesdrop on messages in the aggregation. PPP1 uses the minimum connectivity and computes only a hash function on the user side. Considering these properties, one can conjecture that on the user side, no other protocol that encrypts each measurement can be faster, i.e., the encryption of PPP1 is the fastest possible with a very low lower bound in comparison with previous protocols.

6.3 PPP2 Based on Commitments and ECC

PPP2 allows users to send their signed commitments directly to the counting agent who can detect failures in the communication network. Similar to PPP1, PPP2 assumes that a PMU can provide aggregated measurements a_j. Figure 6.3 depicts the communication model of PPP2.

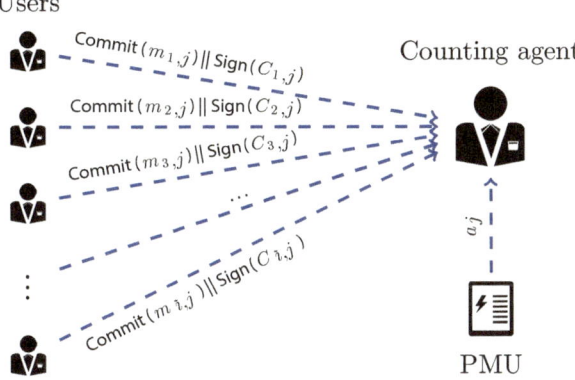

Fig. 6.3 Communication model of PPP2

PPP2 uses ECC for users to create commitments. Since ECC is known for providing short keys and for being efficient, it has been used in PPPs, e.g., [6, 30, 34]. To keep this work as self-contained as possible, Sect. 6.3.1 presents a short review of ECC. A description of its theory is out of scope and may be found in textbooks [14, 22, 26]. The motivations to use elliptic curves may be understood in Sect. 6.3.5. The aim is to present their benefit and how use them in PPP2.

6.3.1 Cryptographic Primitives

Koblitz [27] and Miller [33] independently introduced ECC. The core idea is to use the Discrete Logarithm Problem (DLP) in a group structure on an elliptic curve over a finite field $\mathbb{F}$.

An elliptic curve Ω over a field $\mathbb{F}$ is defined by the Weierstraß equation

$$\Omega : y^2 + a_1 xy + a_3 y = x^3 + a_2 x^2 + a_4 x + a_6, \tag{6.5}$$

where $a_1, a_2, a_3, a_4, a_6 \in \mathbb{F}$ and its discriminant Δ is different from zero, and

$$\begin{cases} \Delta = -d_2^2 d_8 - 8d_4^3 - 27d_6^2 + 9d_2 d_4 d_6 \\ d_2 = a_1^2 + 4a_2 \\ d_4 = 2a_4 + a_1 a_3 \\ d_6 = a_3^2 + 4a_6 \\ d_8 = a_1^2 a_6 + 4a_2 a_6 - a_1 a_3 a_4 + a_2 a_3^2 - a_4^2. \end{cases}$$

An addition operation together with the set Ω of points that satisfy Eq. (6.5) and an identity called point at infinity (∞) form an abelian group $\Omega(\mathbb{F})$, i.e.,

$$\Omega(\mathbb{F}) = \{(x, y) \in \mathbb{F} \times \mathbb{F} \mid y^2 + a_1 xy + a_3 y = x^3 + a_2 x^2 + a_4 x + a_6\} \cup \{\infty\}.$$

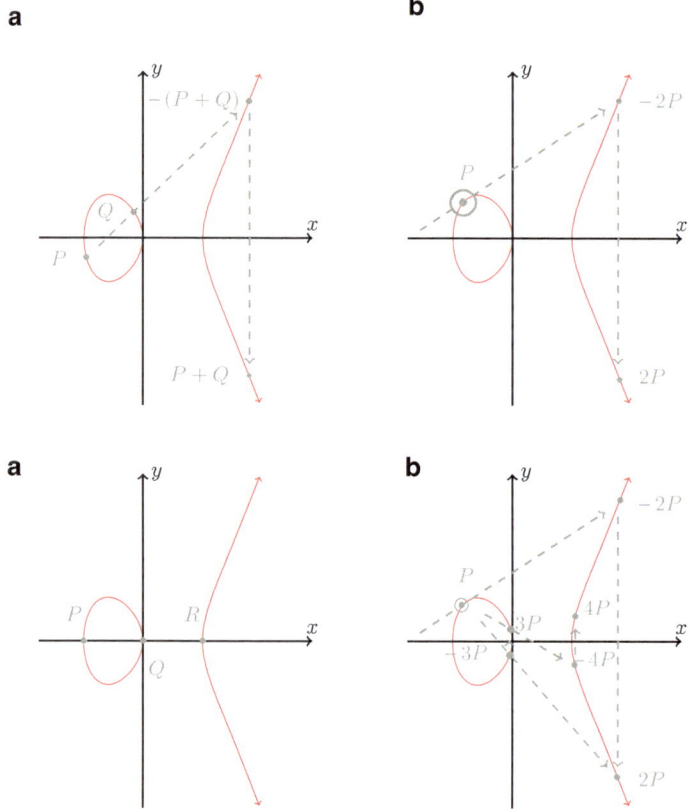

Fig. 6.4 Group structure on an elliptic curve $\Omega(\mathbb{F})$. (**a**) Point addition. (**b**) Point doubling. (**c**) Inflection points. (**d**) Scalar multiplication

This group structure can also be defined geometrically. Figure 6.4 depicts the geometry of an elliptic curve over $\mathbb{R}$. The addition of two points P and Q can be found drawing a straight line through P and Q until the line intercepts another point $-(P + Q)$ whose reflection about the x-axis is $P + Q$, cf. Fig. 6.4a. If $P = Q$, we draw the tangent until it intercepts another point $-2P$ whose reflection is $2P$, cf. Fig. 6.4b. We can also see that the identity is out of the plane, namely $P + Q - (P + Q) = 2P - 2P = P - P = \infty$. Thus, if P is an infection point, then $P + P = P - P = \infty$, cf. Fig. 6.4c. Excluding the inflection points, the others can be used as a trapdoor based on scalar multiplication, cf. Fig. 6.4d. Similarly to modular exponentiation, scalar multiplication can be efficiently computed, cf. Algorithm 6.

In cryptography, elliptic curves are used over finite fields, and point additions are used to compute the scalar multiplications, e.g., $7P = 2(2P) + 2P + P$. Thus, we can compute $R = kP$ given k and P. However, scalar multiplication is a cryptographic trapdoor, i.e., given R and P, it is computationally intractable to find an integer k s.t. $R = kP$. Finding k is known as the Elliptic Curve Discrete Logarithm Problem (ECDLP). Note that this book follows the notation in which uppercase are points in Ω and lowercase are elements of the field $\mathbb{F}$.

The equations from the group law to compute point addition can be deduced from the geometry, and they can be simplified with Eq. (6.5) according to the characteristic of the field $\mathbb{F}$, i.e., according to the smallest number of times necessary for adding the multiplicative identity element to result the additive identity element, e.g., the characteristic of $\mathbb{Z}_3$ is 3, because $1 + 1 + 1 = 0 \mod 3$.

The simulation in Chap. 8 uses the curve P-192 with large characteristic in Appendix B. For curves with characteristic bigger than three, we can simplify and transform Eq. (6.5) to

$$y^2 = x^3 + ax + b,$$

where $a, b \in \mathbb{F}$ and the discriminant is given by

$$\Delta = -16(4a^3 + 27b^2).$$

In this case, the point addition $P + Q = (x_1, y_1) + (x_2, y_2) = (x_3, y_3)$ s.t. $P \neq \pm Q$ is given by

$$(x_3, y_3) = \left(\left(\frac{y_2 - y_1}{x_2 - x_1} \right)^2 - x_1 - x_2, \frac{y_2 - y_1}{x_2 - x_1}(x_1 - x_3) - y_1 \right). \tag{6.6}$$

If $P = -Q$, then $P + Q = \infty$. Thus, the point doubling $2P = (x_1, y_1) + (x_1, y_1) = (x_3, y_3)$ is given by

$$(x_3, y_3) = \left(\left(\frac{3x_1^2 + a}{2y_1} \right)^2 - 2x_1, \frac{3x_1^2 + a}{2y_1}(x_1 - x_3) - y_1 \right). \tag{6.7}$$

With point addition and point doubling, we can write Algorithm 6 for scalar multiplication, which is similar to the modular exponentiation given by Algorithm 16.

As a numerical example, Fig. 6.5 depicts the points of the elliptic curve

$$y^2 = x^3 + x + 6 \tag{6.8}$$

over the integers modulo 19, i.e. $\mathbb{Z}_{19}$. For Eq. (6.8), the discriminant is given by $\Delta = 4 \cdot 1^3 + 27 \cdot 6^2 \mod 19 = 7 \neq 0$. Note that ∞ does not appear in Fig. 6.5, because it is out of the Cartesian plane. The point $P = (0, 5)$ is a generator of the cyclic group generated by the points of Eq. (6.8) and the point at infinity, i.e.,

$$\{kP \mid k \in \mathbb{N}\} = \Omega(\mathbb{Z}_{19}) = \{(x, y) \in \mathbb{Z}_{19} \times \mathbb{Z}_{19} \mid y^2 = x^3 + x + 6\} \cup \{\infty\}.$$

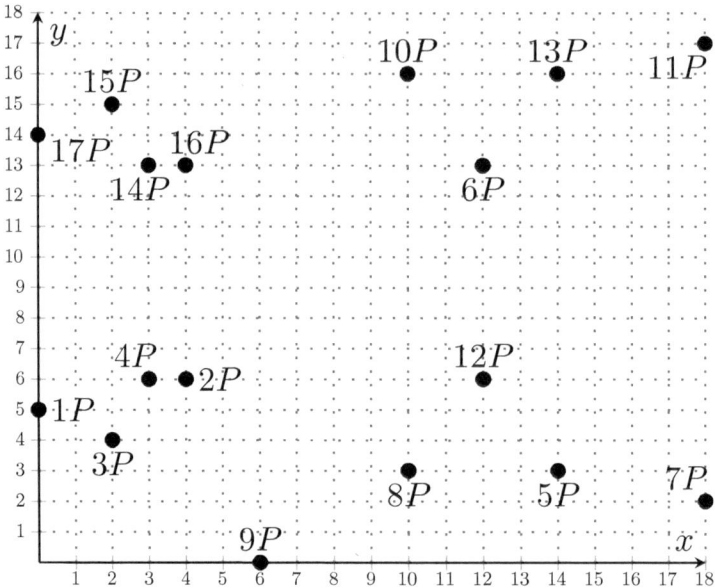

Fig. 6.5 Elliptic curve over $\mathbb{Z}_{19}$

Algorithm 6: Scalar multiplication

Input: An elliptic curve Ω, a point $P \in \Omega$, and an integers k, s.t. $k_t = \sum\limits_{\iota=1}^{l} 2^{\iota-1} e_\iota$, where l is
the bit length of k and $e_\iota \in \{0, 1\}$.
Output: $k \cdot P \in \Omega$.
1   $Q \leftarrow \infty$
2 **for** $\iota = l$ **to** 1 **by** -1 **do**
3     $Q \leftarrow 2 \cdot Q$ // v.s. Eq. (6.7)
4   **if** $e_\iota = 1$ **then**
5        $Q \leftarrow Q + P$ // v.s. Eq. (6.6)

6 **return** Q

Whereas this elliptic curve group $\Omega(\mathbb{Z}_{19})$ is cyclic, we can sort the points using Algorithm 6. We just need to multiply scalars to the generator $P = (0, 5)$. Therefore, $1P = P = (0, 5)$, $2P = (4, 6)$, etc.

Section 6.3.3 shows that Eq. 6.8 is not secure for cryptography. Nevertheless, the points of $\Omega(\mathbb{Z}_{19})$ sorted by Algorithm 6 show us that they are scrambled in the Cartesian coordinates. In Fig. 6.5, the points do not follow a sequence in the plane, e.g.,

$$4P + 12P = 16P \implies (3, 6) + (12, 6) = (4, 13).$$

The point $9P$ is the unique inflection point, and $18P$ is the point at the infinity, i.e. $9P + 9P = 18P = \infty$. All other points have an inverse deferent from itself, e.g., $1P + 17P = 2P + 16P = \infty$.

Trying to compute some points, we can see that they are irregularly spread, and the addition of two points does not indicate the position of the result.

6.3.2   Proposed Protocol

In the set-up phase, the counting agent and users agree on an elliptic curve Ω with secure parameter, cf. Sect. 6.3.3, and on a secure hash function s.t. it behaves as a one-way function and has collision resistance. In addition, they also agree on a base point $P \in \Omega$ with high order, and each user i chooses a permanent key $k_i \in \mathbb{F}$. Using a fully connected SDC-Net or PPP1, the counting agent receives the sum s of the private keys k_i, i.e.,

$$s \overset{\text{def.}}{=} \sum_{i=1}^{\tilde{i}} k_i. \tag{6.9}$$

The hash function H is used to define a hash function H_Ω over $\Omega(\mathbb{F})$ as

$$R_j \overset{\text{def.}}{=} (x, y) \overset{\text{def.}}{=} \mathsf{H}_\Omega(j) \overset{\text{let}}{=} \left(\min \left(\{ r | r \geq \mathsf{H}(j) \text{ and } (r, y) \in \overline{\Omega} \} \right), y \right), \tag{6.10}$$

where $\overline{\Omega}$ is a subset of Ω that contains elements of high order, and x and y are coordinates of a point in the curve.

For each round j, the users can commit and sign their measurements. The commitment function is given by

$$C_{i,j} \overset{\text{def.}}{=} \mathsf{Commit}(m_{i,j}) \overset{\text{let}}{=} k_i \cdot R_j + \mathsf{Value}\left(m_{i,j} \right) \cdot P. \tag{6.11}$$

where $R_j \overset{\text{def.}}{=} \mathsf{H}_\Omega(j)$, s.t. $||$ denotes string concatenation. Algorithm 7 describes the steps followed by the users. PPP2 can use any signature function.

Algorithm 7: PPP2—Commitment

Input: Measurement $m_{i,j}$.
Output: Signed and committed measurement $C_{i,j}$.
1 $C_{i,j} \leftarrow \mathsf{Commit}(m_{i,j})$ // v.s. Eq. (6.11)
2 $\mathsf{S}_{i,j} \leftarrow \mathsf{Sign}\,(C_{i,j})$
3 **return** $C_{i,j} || \mathsf{S}_{i,j}$

After users compute the commitment function, they sign their measurements and send them directly to the counting agent who can verify the bill $b_i^\$$ and also whether the consolidated monetary value $c_j^\$$ is equivalent to the aggregated measurement a_j. As the messages arrive, the counting agent verifies the digital signature $\mathfrak{S}_{i,j}$ and, if they are correct, calculates the aggregation

$$A_j \stackrel{\text{def.}}{=} \sum_{i=1}^{\tilde{i}} C_{i,j}.$$

Thereafter, the counting agent can perform the aggregated measurement verification as well as PPP1. In addition, PPP3 can detect deceptive users and enables billing verification.

6.3.2.1 Privacy-Unfriendly Individual Measurement Verification

The counting agent can verify an individual measurement $m_{i,j}$ of a user i who can just present $m_{i,j}$ and $V \stackrel{\text{def.}}{=} k_i \cdot \mathsf{H}_\Omega(j)$ to the counting agent. Thus, they can compute

$$\mathsf{Open}\left(C_{i,j}, m_{i,j}, V\right) \stackrel{\text{let}}{=} \left(C_{i,j} \stackrel{?}{=} V + \mathsf{Value}\left(m_{i,j}\right) \cdot P\right), \tag{6.12}$$

where $C_{i,j}$ is the commitment of the measurement $m_{i,j}$ sent by the user i to the counting agent in the round j. The commitment can be open **iff** the values in Eq. (6.12) is correct.

6.3.2.2 Aggregated Measurement Verification

With s, the counting agent opens the consolidated monetary value $c_j^\$$ of the commitments calculating

$$\mathsf{Open}\left(A_j, a_j, s \cdot R_j\right) \Leftrightarrow A_j \stackrel{?}{=} s \cdot R_j + c_j^\$ \cdot P. \tag{6.13}$$

Thus,

$$\mathsf{Open}\left(A_j, a_j, s \cdot R_j\right) \Leftrightarrow \sum_{i=1}^{\tilde{i}} C_{i,j} \stackrel{?}{=} \left(\sum_{i=1}^{\tilde{i}} k_i\right) \cdot R_j + c_j^\$ \cdot P \tag{6.14}$$

but Eq. (6.14) holds when

$$c_j^\$ \stackrel{?}{=} \sum_{i=1}^{\tilde{i}} \mathsf{Value}\left(m_{i,j}\right). \tag{6.15}$$

Algorithm 8: PPP2—Aggregated measurement verification

Input: encrypted consolidated consumptions $\mathfrak{C}_j$ and aggregated measurement a_j.
Output: Either Incorrect or Correct.
1 $A_j \leftarrow \infty$
2 $c_j^\$ \leftarrow a_j$
3 **for** $i \leftarrow 1$ **to** $\tilde{\imath}$ **do**
4 $\quad$ **if** Verify $\left(C_{i,j}||\mathfrak{S}_{i,j}\right)$ **then**
5 $\quad\quad$ $\lfloor$ $A_j \leftarrow A_j + C_{i,j}$
6 $\quad$ **else**
7 $\quad\quad$ $|$ Apply policies
8 $\quad\quad$ $\lfloor$ **return** Incorrect

9 **if** Open $\left(A_j, c_j^\$, s \cdot R_j\right)$ **then**
10 $\quad$ $\lfloor$ **return** Correct
11 **else**
12 $\quad$ $|$ Apply policies
13 $\quad$ $\lfloor$ **return** Incorrect

Therefore, the counting agent knows whether the consolidated consumption c_j is correct, because the consolidated monetary value $c_j^\$$ is given by a function of the consolidated consumption c_j that should be approx. the aggregated measurement a_j. If loss of energy is detected—v.s. Sect. 4.1.1—the counting agent can search for the proper value of $c_j^\$$ that opens the commitment in Eq. (6.13). If $c_j^\$$ is too high or too low, the consolidated monetary value is wrong, i.e., the supplier defines the accepted losses. Therefore, if they are correct, the counting agent knows the missing amount. Algorithm 8 describes the process of verifying the aggregated measurement a_j. The function Verify $\left(C_{i,j}||\mathfrak{S}_{i,j}\right)$ returns true if the $C_{i,j}$ matches with its digital signature $\mathfrak{S}_{i,j}$. Note that if a counting agent did not receive a message, the signature is not verified after a period.

6.3.2.3 Detecting Deceptive Users

Suppose that a user inserted a huge value to disrupt the communication in the round j, and that the counting agent does not want to wait for the bill $b_{\tilde{\imath}}^\$$ to detect the deceptive user i, cf. Table 4.1.

The counting agent can detect the source in $\log_2(\tilde{\imath})$ steps, where $\tilde{\imath}$ is the number of users. The counting agent groups the users into two sets $\mathcal{U}_1$ and $\mathcal{U}_2$ and verifies from which set the problem comes. The counting agent can group the users from the set with problems into two new sets again, and can repeat the procedure until the counting agent detects the user. In the first step, users can use a fully connected SDC-Net or PPP1 in order for the set of users $\mathcal{U}_1$ to send

$$v_1 \overset{\text{def.}}{=} \sum_{i \in \mathcal{U}_1} \text{Value}\left(m_{i,j}\right),$$ (6.16)

and

$$V_1 \overset{\text{def.}}{=} \prod_{i \in \mathcal{U}_1} R_j^{k_i}.$$ (6.17)

Since the signed commitments $C_{i,j}$ are known, the counting agent calculates

$$\prod_{i \in \mathcal{U}_1} C_{i,j} \overset{?}{=} v_1 \cdot V_1.$$ (6.18)

If Eq. (6.18) is correct and v_1 is not a huge value, the counting agent requests the other set of users to send

$$v_2 \overset{\text{def.}}{=} \sum_{i \in \mathcal{U}_2} \text{Value}\left(m_{i,j}\right),$$ (6.19)

and

$$V_2 \overset{\text{def.}}{=} \prod_{i \in \mathcal{U}_2} R_j^{k_i}.$$ (6.20)

Similarly, the counting agent computes

$$\prod_{i \in \mathcal{U}_2} C_{i,j} \overset{?}{=} v_2 \cdot V_2.$$ (6.21)

and verifies if Eq. (6.21) is correct and v_2 is not a huge value. Whereas

$$c_j^\$ = v_1 + v_2,$$

one of these values should be huge or one of the two equations should not hold. Therefore, the counting agent knows which set has a problem and can apply the same strategy recursively over the set with a problem. Certainly, the counting agent might learn something about the users, for instance, if a subset has no consumption. To minimize the leakage, users from the set without a problem can be regrouped in the subsets generated by the set with a problem. This strategy generates sub-consolidated consumptions and allows the counting agent to detect the problem source with the same number of steps. Algorithm 9 describes the process of detecting the set with anomalous behavior. Recursive iterations of Algorithm 9 lead to the detection of the deceptive users. For example, in the last interaction of the search for the user that inserted a huge value, the counting agent knows that one of

---

Algorithm 9: PPP2—Detecting deceptive users

Input: v_1 and V_1
`// v.s. Eqs.(6.16), (6.17), (6.20) and (6.20)`
Output: Set with problem.
1 $Q \leftarrow \infty$
2 **foreach** $i \in \mathcal{U}_1$ **do**
3 $\quad \lfloor \ Q \leftarrow Q + C_{i,j}$

4 **if** v_1 *is expected and* $\mathsf{Open}\,(Q, v_1, V_1)$ **then**
5 $\quad |$ **return** $\mathcal{U}_2$

6 **else**
7 $\quad \lfloor$ **return** $\mathcal{U}_1$

---

two users sent the huge value. To verify without a breach of privacy, each of them joins with a disjoint set of users. If one set is verified without the huge value, the user of the other set has inserted it.

6.3.2.4 Billing Verification

Besides verification with the aggregated measurement a_j, the counting agent and a user i can verify the correctness of the bill $b_i^\$$. The account is similar to aggregated measurement verification in Sect. 6.3.2.2. However, billing verification requires neither in-network aggregation nor an SDC-Net.

To verify, the user i presents

$$V_i \overset{\text{def.}}{=} \sum_{j=1}^{\tilde{j}} k_i \cdot \mathsf{H}_\Omega(j), \tag{6.22}$$

and the counting agent calculates

$$B_j = \sum_{j=1}^{\tilde{j}} C_{i,j}$$

and

$$\mathsf{Open}\left(B_i, b_i^\$, V_i\right) \Leftrightarrow B_i \overset{?}{=} V_i + b_i^\$ \cdot P. \tag{6.23}$$

Thus,

$$\mathsf{Open}\left(B_i, b_i^\$, V_i\right) \Leftrightarrow \sum_{j=1}^{\tilde{j}} C_{i,j} \overset{?}{=} k_i \cdot \left(\sum_{j=1}^{\tilde{j}} R_j\right) + b_i^\$ \cdot P, \tag{6.24}$$

Algorithm 10: PPP2—Billing verification

Input: Bill $b_i^\$$ and V_1.
// v.s. Eq. (6.22)
Output: Correctness of bill $b_i^\$$.
1 $B_i \leftarrow \infty$
2 **for** $j \leftarrow 1$ **to** $\tilde{j}$ **do**
3 $\quad \lfloor \; B_i \leftarrow B_i + C_{i,j}$

4 **if** Open $\left(B_i, b_i^\$, V_i \right)$ **then**
5 $\quad \lfloor$ **return** $b_i^\$$ is correct

6 **else**
7 $\quad \lfloor$ **return** $b_i^\$$ is incorrect

---

where $R_j = \mathsf{H}_\Omega(j)$, but Eq. (6.14) holds when

$$b_i^\$ \stackrel{?}{=} \sum_{j=1}^{\tilde{j}} \text{Value}\left(m_{i,j} \right). \tag{6.25}$$

Therefore, the counting agent and each user i can verify whether the bill $b_i^\$$ is correct. Algorithm 10 describes the process of verifying the bill $b_i^\$$.

6.3.3 Security Analysis

The PPP2's security depends on elliptic curve parameters and the attacker model, namely, a dishonest user and the information that the counting agent can get. Thus, the security analysis is divided into two parts, namely, selection of secure parameters and attacker model.

The parameter selection for ECC is more complicated than for cryptographic schemes based on Integer Factorization Problem (IFP). However, it is harder to solve the DLP over ECC than to solve the DLP over a finite group $(G, \circledast)$ of integers $\mathbb{Z}$. Therefore, the size of the elliptic curve group can be considerably smaller than the size of $(G, \circledast)$, cf. Sect. 6.3.5.

Menezes et al. [32] presented an algorithm to solve the DLP over supersingular elliptic curves with complexity sub-exponential and Smart [44] presented an algorithm to solve the DLP over prime-field anomalous elliptic curves with polynomial complexity. A curve Ω is supersingular over a finite field $\mathbb{F}$ **iff** the trace of Frobenius t is zero, i.e., $t \equiv 0 \mod p$. Since t is defined by Hasse's theorem $\#\Omega(\mathbb{F}_q) = q + 1 - t$, where $|t| \leqslant 2\sqrt{q}$, then supersingular curves generate groups with $q + 1$ elements, i.e., $\#\Omega(\mathbb{F}_q) = q + 1$. A curve Ω is prime-field anomalous **iff** $t = 1$, thus $\#\Omega(\mathbb{F}_p) = p$. Schoof [42] presented an algorithm that determines the order of an elliptic curve group $\#\Omega(\mathbb{F}_q)$ with logarithmic time $O(\log^9 q)$, i.e., polynomial time with respect to the bit-length of q. Thus, we can determine the group size, and therefore, if the curve is a supersingular or a prime-field anomalous elliptic curve. Another factor that weakens the ECDLP is the group structure. Specifically, if $P = (x, y)$ is a based point belonging to $\Omega(\mathbb{F}_q)$ and generating a cyclic subgroup $\langle G \rangle \subseteq \Omega(\mathbb{F}_q)$ s.t. $\langle G \rangle = \{kP : k \in \mathbb{Z}\}$, then the security of the ECDLP is determined by

$$h = \frac{\#\Omega(\mathbb{F}_q)}{\#\langle G \rangle}.$$

The smaller h is, the better. If $h = 1$, $\Omega(\mathbb{F}_q)$ is a cyclic group. For $h \leqslant 4$, we say that $\Omega(\mathbb{F}_q)$ is *almost cyclic*. Therefore, we have three core factors to verify, namely:

- $\#\Omega(\mathbb{F}_p) \neq p$ excludes prime-field anomalous curves;
- $t \not\equiv 0 \pmod{p}$ excludes supersingular curves;
- $h \leqslant 4$ excludes small subgroups of P.

Since the secure parameters are established, we can discuss what the attacker can do. The set-up phase depends on the security of PPP1 or an SDC-Net. In contrast to PPP1, the attacker might intercept the messages and try to recover k_i or $m_{i,j}$ from the commitment function defined in Eq. (6.11). However, this is infeasible for secure parameters. Since messages are signed, the attacker cannot compromise them and users cannot repudiate them. The open function returns true **iff** the parameters are correct. If users signed wrong messages, the counting agent can discover them by re-aggregating the commitments or by awaiting the bill $b_i^\$$, which can also be verified.

Note that the counting agent can search small values of consolidated monetary value $c_j^\$$, which are close to aggregated measurement a_j, but no one can search large values like k_i. More details can be found in Sect. 6.3.5 and Chap. 8.

6.3.4 Privacy Analysis

To keep privacy, the individual measurements should be protected. The counting agent can verify the consolidated consumption c_j, consolidated monetary value $c_j^\$$, and bill $b_i^\$$ **iff** equations hold in Algorithms 8–10, respectively.

Measurements from the same user i or the same round j cannot be related because of the hash function H.

Users might collude, but we should assume that at least 2 users are honest. Since s in Eq. (6.9) is the sum of all keys, $\tilde{\imath} - 1$ users need to collude to disclose the key of one user. Without disclosing the key k_i, users cannot read an individual measurement $m_{i,j}$ from the user i. Therefore, the collusion of $\tilde{\imath} - 2$ users is not enough to disclose one key.

Note that PPP2 has much weaker assumptions than PPP1, e.g., the counting agent could collude with 2 users—2 is enough—to leak individual encrypted measurements $\mathfrak{M}_{i,j}$ with PPP1 from a user i. Moreover, PPP1 requires the honest-but-curious assumption and that the attacker cannot have access to the aggregation process, whereas PPP2 does not require such assumptions.

6.3.5 Performance Analysis

The National Institute of Standards and Technology (NIST) in its FIPS 186-2 recommends some elliptic curves, Appendix B presents one of them. Scalar multiplication for ECC is well known to be faster than modular exponentiation used in schemes based on IFP. This section presents a time complexity analysis. Details about processing time may be found in the simulation presented in Chap. 8. Molina-Markham et al. [34] already showed the feasibility of running ECC on smart meters.

Scalar multiplication usually works in smaller numeric sets than modular exponentiation with the equivalent level of security. Thus, ECC is known to have smaller keys. Certainly, the set size and the key length depend on the best algorithms to find the key. In the literature, the fastest algorithm to solve the IFP [22] asymptotically has complexity

$$\exp\left(\left(\left(\frac{64}{9}\right)^{1/3} + O(1)\right)(\ln n)^{1/3}(\ln \ln n)^{2/3}\right), \tag{6.26}$$

where n is the product of two safe primes. Such an algorithm is known as general number field sieve (GNFS). For integers of the form $\alpha^{\beta} + \gamma$ where α and γ are small integers, the special number field sieve (SNFS) is faster than the GNFS. The SNFS reduces the 64 in the numerator of Eq. (6.26) to 32. In contrast, the fastest algorithm found in the literature to solve the DLP, and thus, ECDLP [22] has complexity

$$\sqrt{\frac{\pi o}{2}}, \tag{6.27}$$

where o is the order of P. The key can be reduced, because the time complexity of the ECDLP in Eq. (6.27) is lower than the time complexity of the IFP in Eq. (6.26). Using (6.26) and (6.27), we can construct Table 6.1 to compare the effort of both algorithms with the effort required by brute force attack. Algorithm 18 in Appendix A shows how to construct the columns DLP, GNFS, and SNFS. The values in the column NIST are recommended by NIST in the Special Publication 800-57–Part 1 (Revision 3–July 2012).

Consequently, we can reduce the key size when its security is based on the DLP or ECDLP. Moreover, the key size grows much slower for ECDLP than for IFP. Therefore, protocols based on ECDLP tend to be exponentially faster than

Table 6.1 Comparison
between brute force and
minimum key length

Brute force	DLP	GNFS	NIST	SNFS
80	160	851	1,024	1,449
112	224	1853	2,048	3,199
128	256	2538	3,072	4,403
192	384	6707	7,680	11,787
256	512	13,547	15,360	24,000

Fig. 6.6 Comparison
between brute force and
minimum key length

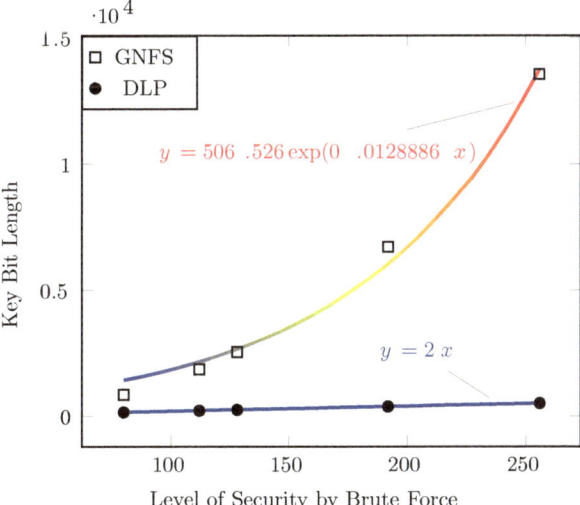

PPPs based on IFP. Since PPP2 uses only two scalar multiplications to commit in Eq. (6.11), PPP2 is not only fast, but also increasingly faster than many other protocols for smart grid, cf. Chap. 3.

With Table 6.1, we can plot the point to obtain a better visualization. Moreover, we can use the brute force as reference—i.e., x-axis—thus, the points generated by IFP can be fitted by the curve $y = 2x$, but the points generated by ECDLP can be fitted by the exponential curve $y = 506.526 \exp(0.0128886x)$. Figure 6.6 depicts the points with their fitted curves.

6.4 PPP3 Based on Asymmetric DC-Nets

PPP3 can provide the same information as PPP1 and PPP2, but its communication model is simpler than PPP1's communication model. Figure 6.7 depicts the communication model used in PPP3. This is similar to PPP2 but PPP3 returns the decrypted consolidated monetary value $c_j^\$$ instead of commitment verification. Moreover, PPP3 also provides verification as done with commitments.

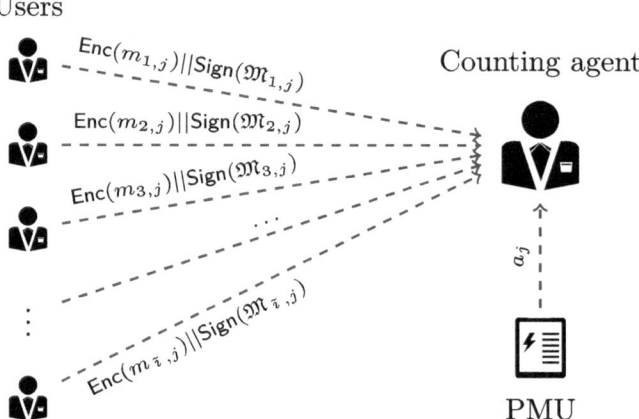

Fig. 6.7 Communication model of PPP3

This protocol uses a technique called ADC-Nets [9]. Specifically, PPP3 uses a fully connected ADC-Net that is equivalent to a complete graph. Section 6.4.1 presents the concept of ADC-Nets and their properties. Section 6.4.2 describes the attacker model. Section 6.4.3 introduces the ADC-Net for smart grids. Section 6.4.4 introduces the verification processes. Section 6.4.5 discusses the security issues, and Sect. 6.4.6 talks about privacy issues. Section 6.4.7 shows that PPP3 is efficient.

6.4.1 Cryptographic Primitives

Before the introduction of ADC-Nets [9], the DC-Nets introduced by Chaum [13] were classified as symmetric [3]. ADC-Nets are defined by properties and have other derived properties.

6.4.1.1 Properties from the Definition

ADC-Nets can be created with many cryptographic primitives. PPP3 uses an ADC-Net [9], which runs over integers. An ADC-Net protocol is defined by the following properties:

1. all properties of SDC-Nets, with the exception of unconditional security;
2. security based on a cryptographic trapdoor function;
3. users can use permanent keys;
4. processing time has complexity at most polynomial;
5. non-iteration over the number of users, with exception of aggregation;
6. users send the minimum number of messages;

7. users can sign their messages;
8. similar to commitments, users can prove that their messages are sent correct.

With this definition, we can see an example of ADC-Net. To construct an ADC-Net, assume that each user i has a private key k_i, a fixed g s.t. $g \in \mathbb{Z}_{n^2}$, and a product of two secret primes n s.t. n is generated by the users [4]. Then, a family of functions that defines the encryption is given by

$$
\mathsf{Enc} : \mathbb{Z}_n \to \mathbb{Z}_{n^2}
$$
$$
\mathsf{Enc}_i(m_{i,j}) \mapsto (1+n)^{m_{i,j}} \cdot g^{h_j \cdot k_i} \mod n^2, \tag{6.28}
$$

where $h_j \overset{\text{def.}}{=} \mathsf{H}(j)$ with H being a secure hash function s.t. it behaves as a one-way function and has collision resistance. Thus, the decryption function is given by

$$
\mathsf{Dec} : \mathbb{Z}_{n^2} \to \mathbb{Z}_n
$$
$$
\mathsf{Dec}\left(\mathfrak{C}_j\right) \mapsto \frac{\left(\mathfrak{C}_j \cdot g^{-h_t \cdot s} \mod n^2\right) - 1}{n}, \tag{6.29}
$$

where $s \overset{\text{def.}}{=} \sum_i^{\tilde{\iota}} k_i$.

Let us verify whether these functions satisfy the properties given in the definition.

Property 1 is strong but can be verified easily. The properties of SDC-Nets come from the cancellation of their key, cf. Sect. 3.2.2. The keys k_i are canceled when their sum is equal to n. Two users 1 and 2 can allow that the sum of their measurements be decrypted independent of the other measurements. From the sum, we need the collusion of $\tilde{\iota} - 1$ to leak the key of one user. Indeed, n in Eqs. (6.28) and (6.29) works as zero in Eq. (3.5). The process is non-iterative, i.e., users can send only a broadcast or send their signed messages directly to a counting agent, but they do not need to rely on a trusted third party (TTP).

Property 2 requires a cryptographic trapdoor function. We can see the mathematical problem in Eq. (6.28), i.e., solve $g^{h_j \cdot k_i} \mod n^2$. Thus, the security is based on the assumption that the DLP over $\mathbb{Z}_{n^2}$ is intractable. This assumption has been used in many schemes, e.g., [37].

Property 3 protects the PPP against overhead to exchange new keys. It is satisfied because the k_i for all users i cannot be related in many encrypted measurement due to the hash function H.

Property 4 enables encrypt and decrypt in the maximum of polynomial time complexity. The most expensive operation in the process is modular exponentiation that can be computed with logarithmic time with respect to its exponent.

Property 5 requires that the encryption and the decryption functions be free of iteration over the number of users. This is the case for Eqs. (6.28) and (6.29). The aggregation is the trivial case, where the measurements of all users should be counted.

Property 6 ensures communication efficiency, i.e., the minimum number of messages sent per measurement $m_{i,j}$ is one. Using the encryption and the decryption functions in Eqs. (6.28) and (6.29), users can send the minimum number of messages.

Property 7 introduces non-repudiation, i.e., no user can deny the authorship of any measurement $m_{i,j}$. Since Property 1 allows a broadcast, the messages can be signed. In fact, Property 7 is also satisfied in an SDC-Net, but it is an explicit property of the definition to ensure a secure commitment in Property 8.

Property 8 ensures verifiability in PPPs. A user i can send v s.t. $v = g^{h_j \cdot k_i}$ and the receiver can verify the measurement $m_{i,j}$ without discovery k_i. Verification of individual measurements is a privacy-unfriendly process in ADC-Net schemes as well as commitment schemes. Users should send v s.t. $v = \prod g^{h_j \cdot k_i}$ and the sum of many measurements $\sum m_{i,j}$.

After the verification of the properties, we have an example of an ADC-Net given by the encryption and the decryption functions in Eqs (6.28) and (6.29). In the following, let us verify more two interesting properties derived from ADC-Net.

6.4.1.2   Derived Properties

The first statement is: Paillier is a particular case of ADC-Net. To verify this statement, let us consider a more complicated ADC-Net with the encryption function given by

$$\mathsf{Enc} : \mathbb{Z}_n \times \mathbb{Z}_n \times \mathbb{Z}_n \to \mathbb{Z}_{n^2}$$
$$\mathsf{Enc}_i(m_{i,j}, k_i, r_{i,j}) \mapsto (1+p)^{m_{i,j}} \cdot h_j^{k_i} r^n \mod n^2, \tag{6.30}$$

where n is the product of two safe primes, $h_j \in \mathbb{Z}_{n^2}$ s.t. $h_j > n$ and $h_j \stackrel{\text{def.}}{=} \mathsf{H}(j)$, and H is a secure hash function s.t. it behaves as a one-way function and has collision resistance. Thus, the decryption function is given by

$$\mathsf{Dec} : \mathbb{Z}_{n^2} \to \mathbb{Z}_n$$
$$\mathsf{Dec}(\mathfrak{C}_j) \mapsto \mathsf{L}((\mathfrak{C}_j \cdot h_2^{n-s})^{\lambda} \mod n^2) \cdot d \mod n, \tag{6.31}$$

where $d \stackrel{\text{def.}}{=} \mathsf{L}(g^{\lambda} \mod n^2)^{-1}$, $\mathfrak{C}_j \stackrel{\text{def.}}{=} \mathsf{Enc}(c_j)$, c_j is the consolidated consumption, and

$$s \stackrel{\text{def.}}{=} \sum_{i=1}^{\tilde{i}} k_i.$$

Equations (6.30) and (6.31) are similar to Paillier's equations given by Eqs. (3.2) and (3.4).

If we give up some ADC-Net properties and make $k_i = 0$ for all user i, the result of Eq. (6.30) is equal to Eq. (3.2) and the result of Eq. (6.31) is equal to Eq. (3.4). Therefore, Paillier encryption and decryption functions are particular cases of an ADC-Net. Other schemes that satisfy the ADC-Net properties can be reduced to Paillier, e.g., [7].

The second statement is: all AHEPs are particular cases of ADC-Nets. To verify this statement, let us only consider AHEPs that are probabilistic encryption schemes with at most polynomial processing time complexity. The creation of AHEPs that are not probabilistic encryption enables attacker to relate the encrypted measurements, cf. Scct. 2.2.2.3. AHEPs that have higher complexity are not scalable. Thus, we cannot say that PPP2 has an AHEP in Eq. (6.11). Therefore, PPP2 is not based on an ADC-Net.

From the probabilistic encryption property, we have

$$\mathsf{Dec}(\mathsf{Enc}_{r_1}(m_{i,j})\mathsf{Enc}_{r_2}(m_{i,j})) = 2m_{i,j}$$

but

$$\mathsf{Enc}_{r_1}(m_{i,j}) \neq \mathsf{Enc}_{r_2}(m_{i,j}).$$

This happens because there is a special number n that cancels the random numbers r_i, e.g., $w = r_i^n \mod n^2$ remembering that

$$\forall\, w \in \mathbb{Z}_{n^2}^*, \ \begin{cases} w^\lambda \equiv 1 \mod n \\ w^{n\lambda} \equiv 1 \mod n^2, \end{cases}$$

where λ is Carmichael's function $\lambda \stackrel{\text{def.}}{=} \lambda(n) = \mathsf{lcm}(p-1, q-1)$ [19]. Thus, we can split up n into many keys s.t.

$$r \odot n = r \odot (k_1 \oplus k_2 \oplus \cdots \oplus k_{\tilde{i}}).$$

We can use a hash function H as a cryptographically secure pseudorandom number generator (CSPRNG). Hence, we can use the hash function H with $h_j \stackrel{\text{def.}}{=} \mathsf{H}(j)$ and the cryptographic trapdoor s.t.

$$v \stackrel{\text{def.}}{=} h_j \odot k_i$$

works as a verifier for a commitment. For security reasons, the round j must never repeat. Since it is a temporal sequence of events by definition of round, each round j has a different value of j. After the aggregation in the round j, we have

$$h_j \odot k_1 \otimes h_j \odot k_2 \otimes \cdots \otimes h_j \otimes k_{\tilde{i}} = h_j \otimes \left(\bigoplus_{i=1}^{\tilde{i}} k_i \right) = h_j \otimes n.$$

Therefore, splitting up a special number, we can construct an ADC-Net for each AHEP.

6.4.1.3 ADC-Nets and Their Applications

ADC-Nets are a generalization of AHEPs. However, the former can be much more efficient than the latter, because a randomized number can be split. The latter is normally called homomorphic encryption, because the primitives can be used to derive other operations, for instance, multiplication. The former can also be used to compute other operations. However, the operation should be applied to all measurements. Each AHEP is a particular case of an ADC-Net. In addition, this result might also be true for fully homomorphic encryption [20], which can compute all circuits with encrypted measurements. Research in fully homomorphic encryption is promising but still in development and too expensive for smart meters.

Practical homomorphic encryption schemes have been used in different levels of applications. Examples of these are protocols for e-voting [15], reputation systems [25], trust [17], sensor networks [39], multi-party computation [16], e-cash [12], mobile sensing [29], image processing [46], and smart grids [41]. Indeed, AHEPs may be applied in many other applications that require protection of privacy.

Similar to AHEP, SDC-Nets are applied in many scenarios. However, SDC-Nets as well as ADC-Nets can enforce privacy. Thus, the respective PPPs should be adapted to this enforcement, but this is not an obligation, because a star ADC-Net can be set in a similar way that it is used in PPP1 to behave like an AHEP. Indeed, ADC-Nets as well as SDC-Nets can behave as an AHEP. Therefore, all PPPs that use an AHEP can be updated with ADC-Nets.

Solutions based on AHEPs enforce neither privacy by means of aggregation nor security, because the private key might be compromised. In fact, some institutions—and perhaps, attackers—can decrypt individual measurements. In contrast, SDC-Nets or ADC-Nets can enforce security and privacy, which can only be kept if the aggregations can only be decrypted with the participation of all users. Thus, individual measurements cannot be decrypted. In addition, they are still fault tolerant, because users send their measurements directly to their counting agent, which can detect problems in the communication channel and can re-initialize the protocols for the correct users. The unique information lost belongs to the fault measurements, but the protocols continue to work. To maintain security, the leakage of some keys is not enough to compromise PPPs based on SDC-Nets and ADC-Nets. They are compromised with the leakage of all keys but one. Theoretically, SDC-Nets can ensure perfect secrecy using truly random keys only once, as a Vigenère-Vernam-Shannon scheme, better known as the One-time pad.

It is necessary to highlight that the PPPs should be updated to enforce privacy. Without enforcement by cryptography, the right to privacy will likely be violated.

6.4.2 Attacker Model

PPP3 assumes the possibility that an attacker might access and control some meters. Equivalently, we can say that many users might have malicious behavior. From the DC-Nets property, the collusion of $\tilde{\imath} - 1$ users is necessary to disclose data from one user. Thus, if 2 users are honest, their keys cannot be disclosed. However, their privacy has a high risk of leakage because the consolidated monetary value only has the aggregation of two measurements. Therefore, the number of honest users determines the number of measurements aggregated to protect the customers' privacy. Regarding the trust model, the users are considered malicious, but a number of users should behave honestly, cf. Chap. 5.

6.4.3 Proposed Protocol

In the set-up phase, the users and their counting agents agree on a product of primes n, for instance as given in [4]. Each user i chooses a private key k_i, the counting agent chooses a private key k_0, and subsequently, they determine

$$s \stackrel{\text{def.}}{=} \sum_{i=0}^{\tilde{i}} k_i$$

with an SDC-Net in a way that the counting agents know s. They can use PPP1 or a fully connected SDC-Net. With the initial parameters, users can compute the encryption function defined by

$$\begin{aligned} \mathsf{Enc} &: \mathbb{Z}_n \to \mathbb{Z}_{n^2} \\ \mathsf{Enc}_i(m_{i,j}) &\mapsto (1+n)^{\mathrm{Value}(m_{i,j})} \cdot g^{h_j+k_i} \mod n^2, \end{aligned} \tag{6.32}$$

where $h_j \stackrel{\text{def.}}{=} \mathsf{H}(j)$ with H being a secure hash function s.t. it behaves as a one-way function and has collision resistance.

Each user i encrypts the measurement $m_{i,j}$ with Algorithm 11 and sends the result $\mathfrak{M}_{i,j}$ directly to the counting agent. Hence, if a message does not come on time or the signature does not match, the counting agent knows whom to request the message to be re-sent. Indeed, the counting agent can apply any policies. For example, it can send an employee to verify user's meter or sets up the protocols again, excluding the user.

Algorithm 11: PPP3—Encryption

Input: Measurement $m_{i,j}$.
Output: Signed and encrypted measurement $\mathfrak{M}_{i,j}$.
1 $\mathfrak{M}_{i,j} \leftarrow \mathsf{Enc}\,(m_{i,j})$ // v.s. Eq. (6.32)
2 $\mathfrak{S}_{i,j} \leftarrow \mathsf{Sign}\,(\mathfrak{M}_{i,j})$
3 **return** $\mathfrak{M}_{i,j} \| \mathfrak{S}_{i,j}$

Algorithm 12: PPP3—Aggregation and decryption

Input: Encrypted measurements $\mathfrak{M}_{i,j}$ and their digital signature $\mathfrak{S}_{i,j}$.

Output: Consolidated monetary value $c_j^\$$.

1 $\mathfrak{C}_j \leftarrow 1$
2 **for** $j \leftarrow 1$ **to** $\tilde{j}$ **do**
3   **if** Verify $(\mathfrak{M}_{i,j} \| \mathfrak{S}_{i,j})$ **then**
4        $\lfloor$ $\mathfrak{C}_j \leftarrow \mathfrak{C}_j \cdot \mathfrak{M}_{i,j} \mod n^2$

5   **else**
6        $\lfloor$ Apply policies

7 $c_j^\$ \leftarrow \mathsf{Dec}(\mathfrak{C}_j)$ // v.s. Eq. (6.34)
8 **return** $c_j^\$$

After the encryption of the measurements, the aggregation is given by

$$\mathfrak{C}_j \overset{\text{def.}}{=} \prod_{i=1}^{\tilde{i}} \mathfrak{M}_{i,j} \mod n^2, \tag{6.33}$$

and the description function is given by

$$\mathsf{Dec} : \mathbb{Z}_{n^2} \to \mathbb{Z}_n$$

$$\mathsf{Dec}\left(\mathfrak{C}_j\right) \mapsto \frac{\left(\mathfrak{C}_j \cdot g^{-\tilde{i} \cdot h_t - s + k_0} \mod n^2\right) - 1}{n}, \tag{6.34}$$

where $s \overset{\text{def.}}{=} \sum_{i=0}^{\tilde{i}} k_i$.

The counting agent can aggregate with Algorithm 12 when the messages are arriving. However, Algorithm 12 only returns the consolidated monetary value after all encrypted measurements were aggregated.

Algorithm 12 might be split into two: one for aggregation and the other for description. In this case, the counting agent can outsource the aggregation process and runs only Line 7 to decrypt the consolidated monetary value.

6.4.4   Verification Property

In contrast to the other PPPs presented in this chapter, PPP3 enables users and their counting agent to verify the consolidated monetary value and bill with decryption. PPP3 can compute everything that PPP2 computes. In addition, PPP3 can decrypt the consolidated monetary value $c_j^\$$. In fact, the verifications work as commitment schemes do [38]. Hence, it is even possible to verify single measurements, but this is not a privacy-friendly procedure.

6.4.4.1 Privacy-Unfriendly Individual Measurement Verification

ADC-Nets enable verification similar to commitment schemes. Hence, we can define a process of verifying the values as done in open functions for commitment schemes and PPP2. For the sake of simplicity, the verification is presented without such open functions. Thus, this section presents the equations to verify the encrypted measurements.

To verify an individual measurement $m_{i,j}$, a user i just sends $m_{i,j}$ and $v \overset{\text{def.}}{=} g^{h_j + k_i}$ mod n^2 to the counting agent that computes

$$\mathfrak{M}_{i,j} \overset{?}{=} (1+n)^{\text{Value}(m_{i,j})} \cdot v \mod n^2, \tag{6.35}$$

where $\mathfrak{M}_{i,j}$ is the encrypted measurement that was previously sent by the user. They can verify whether Eq. (6.35) is correct. If it is incorrect, then the encrypted measurement $\mathfrak{M}_{i,j}$ corresponds to the measurement $m_{i,j}$ and v. Otherwise, the presented measurement does not match the encrypted measurement and v. Therefore, Eq. (6.35) holds **iff** its values are correct.

6.4.4.2 Aggregated Measurement Verification

In case of the virtualization of a supply network—cf. Sect. 4.1.2—the counting agents may need to prove that the consolidated monetary value is correct. Toward this aim, they show that

$$\mathfrak{C}_j \overset{?}{=} (1+n)^{c_j^\$} \cdot v \mod n^2, \tag{6.36}$$

where $v \overset{\text{def.}}{=} g^{-\bar{\imath} \cdot h_j - s + k_0}$. Then, anyone can verify whether the values are correct. Algorithm 13 describes aggregated measurement a_j verification for PPP3.

6.4.4.3 Detecting Failures and Deceptive Users

Since the messages are signed, the counting agent knows their sender. Thus, if the digital signature $\mathfrak{S}_{i,j}$ is verified and the consolidated monetary value $c_j^\$$ has an unexpected value, a meter sent a wrong measurement $m_{i,j}$ due to failures or an exploited vulnerability. For example, an energy supplier can detect that c_j does not match values provided by a PMU. In this case, the counting agent can discover the respective user during the billing process; cf. Table 4.1.

Algorithm 13: PPP3—Aggregated measurement verification

Input: encrypted measurements $\mathfrak{M}_{i,j}$ with their digital signature $\mathfrak{S}_{i,j}$ and aggregated measurement a_j.

Output: Either Incorrect or Correct.

1 $\mathfrak{C}_j \leftarrow 1$
2 **for** $i \leftarrow 1$ **to** $\tilde{\imath}$ **do**
3   **if** Verify $(C_{i,j}||\mathfrak{S}_{i,j})$ **then**
4          $\mathfrak{C}_j \leftarrow \mathfrak{C}_j \cdot \mathfrak{M}_{i,j} \mod n^2$
5   **else**
6   Apply policies
7   **return** Incorrect
8      $c_j^{\$} \leftarrow \mathsf{Dec}\left(\mathfrak{C}_j\right)$
9 **if** $c_j^{\$} \approx \mathsf{Value}\left(a_j\right)$ **then**
10   **return** Correct
11 **else**
12   Apply policies
13   **return** Incorrect

Suppose that a user i sent a huge measurement $m_{i,j}$ to disrupt the communication. Hence, the counting agent can detect the sender in $\log_2(\tilde{\imath})$ steps, where $\tilde{\imath}$ is the number of users. Similar as in PPP2, the counting agent groups the users in two sets $\mathcal{U}_1$ and $\mathcal{U}_2$ and verifies if the user belongs to $\mathcal{U}_1$ or to $\mathcal{U}_2$. The set with a problem can be re-grouped in an iterative process until the counting agent detects the user. Then, using PPP1 or a fully connected SDC-Net, the first set of users sends

$$v_1 \overset{\text{def.}}{=} \sum_{i \in \mathcal{U}_1} \mathsf{Value}\left(m_{i,j}\right), \tag{6.37}$$

and

$$v_2 \overset{\text{def.}}{=} \prod_{i \in \mathcal{U}_1} g^{h_j + k_i} \mod n^2. \tag{6.38}$$

With the encrypted measurements $\mathfrak{M}_{i,j}$, the counting agent computes

$$\prod_{i \in \mathcal{U}_1} \mathfrak{M}_{i,j} \overset{?}{=} (1 + n)^{v_1} \cdot v_2 \mod n^2. \tag{6.39}$$

If Eq. (6.39) is correct and v_1 is an expected value, the deceptive user belongs to $\mathcal{U}_2$, otherwise, to $\mathcal{U}_1$. Algorithm 14 describes the process of detecting if the user belongs to $\mathcal{U}_1$ or to $\mathcal{U}_2$. The users can be rearranged in two new sets to reduce the number of users in the subset with a problem.

Algorithm 14: PPP3—Detecting deceptive users

Input: $\mathcal{U}_1, \mathcal{U}_2, v_1$ and v_2
`// v.s. Eqs. (6.37) and (6.38)`
Output: Set with problem.
1 $r \leftarrow 1$
2 **foreach** $i \in \mathcal{M}_1$ **do**
3 $\quad \lfloor \; r \leftarrow r \cdot \mathfrak{M}_{i,j} \mod n^2$
4 **if** v_1 *is expected and* $r = (1+n)^{v_1} \cdot v_2 \mod n^2$ **then**
5 $\quad \lfloor \; \text{return } \mathcal{U}_2$

6 **else**
7 $\quad \lfloor \; \text{return } \mathcal{U}_1$

6.4.4.4   Billing Verification

In the billing verification process, the user i sends the bill $b_i^\$$ and v to the counting agent, where

$$b_i^\$ \stackrel{\text{def.}}{=} \sum_{j=1}^{\tilde{j}} \text{Value}\left(m_{i,j}\right)$$

and

$$v_i \stackrel{\text{def.}}{=} \prod_{j=1}^{\tilde{j}} g^{h_j + k_i} \mod n^2. \tag{6.40}$$

The counting agent already has the encrypted measurements $\mathfrak{M}_{i,j}$ signed by the user i, hence it is enough to compute

$$\mathfrak{B}_i \stackrel{\text{def.}}{=} \prod_{j=1}^{\tilde{j}} \mathfrak{M}_{i,j} \mod n^2, \tag{6.41}$$

$$\mathfrak{B}_i \stackrel{?}{=} (1+n)^{b_i^\$} \cdot v_i \mod n^2, \tag{6.42}$$

and to verify whether Eq. (6.42) is correct or not. Algorithm 15 describes the billing verification process.

Algorithm 15: PPP3—Billing verification

Input: Bill $b_i^{\$}$, encrypted measurements $\mathfrak{M}_{i,j}$, and V_i.
    // v.s. Eq. (6.40)
Output: Correctness of bill $b_i^{\$}$.

1 $v_i \leftarrow 1$
2 **for** $j \leftarrow 1$ **to** $\tilde{j}$ **do**
3     $\lfloor \quad v_i \leftarrow v_i \cdot \mathfrak{M}_{i,j} \mod n^2$

4 **if** $\mathfrak{B}_i = (1+n)^{b_i^{\$}} \cdot v_i \mod n^2$ **then**
5     $\lfloor$ **return** $b_i^{\$}$ is correct

6 **else**
7     $\lfloor$ **return** $b_i^{\$}$ is incorrect

6.4.5 Security Analysis

PPP2 and PPP3 are very similar in their properties and features, but in contrast to PPP2, PPP3 can decrypt encrypted consolidated consumption $\mathfrak{C}_j$. Moreover, PPP3 is based on an ADC-Net, which ensures several interesting properties, e.g., users might send one signed message directly to their counting agent, all the computations can be verified, TTP is not necessary, etc. The ADC-Net used in PPP3 is based on the DLP over integers $\mathbb{Z}_n$ and IFP, i.e., its security depends on the assumption that it is intractable to find a key k_i given $g^{h_j + k_i} \mod n^2$, where $g \in \mathbb{Z}_{n^2}$, n is a product of hidden primes [4], $h_j \overset{\text{def.}}{=} \mathsf{H}(j)$, and H is a secure hash function s.t. it behaves as a one-way function and has collision resistance. The assumption that the DLP over integers $\mathbb{Z}_n$ is intractable has been used in other cryptographic schemes, e.g., [37].

Based on the DLP, Algorithm 11 encrypts the measurement $m_{i,j}$ in a way that only the owner of k_i can decrypt $m_{i,j}$. Collusion can disclose a key k_i **iff** $\tilde{i} - 1$ users collude. Algorithm 12 can decrypt the encrypted measurement $\mathfrak{M}_{i,j}$ **iff** all measurements in the round j were aggregated. Algorithm 13 returns "Correct" **iff** the aggregated measurement a_j is correct. Algorithm 14 can be used iteratively to determine who sent a signed encrypted measurement $\mathfrak{M}_{i,j}$ with a wrong value. Algorithm 15 return "bill $b_i^{\$}$ is correct" **iff** the bill $b_i^{\$}$ is correct.

6.4.6 Privacy Analysis

Similar to PPP1 and PPP2, the privacy in PPP3 depends on the negligible probability that a secure hash function returns the same $h_j \overset{\text{def.}}{=} \mathsf{H}(j)$ for different rounds j. If we do not consider the value function and $\mathsf{H}(\alpha) = \mathsf{H}(\beta)$ for some α and β, then the encrypted measurements can be related s.t. $\mathfrak{M}_{i,\alpha} = \mathfrak{M}_{i,\beta}$ **iff** $m_{i,\alpha} = m_{i,\beta}$. If $m_{i,\alpha} \neq m_{i,\beta}$, then $\mathfrak{M}_{i,\alpha} \cdot \mathfrak{M}_{i,\beta}^{-1} = (1+n)^{m_{i,\alpha} - m_{i,\beta}}$. Thus, an attacker can know $m_{i,\alpha} - m_{i,\beta}$ if the hash function returns values with collision. Since we assume a secure hash function s.t. it behaves as a one-way function and has collision resistance, it returns different hashes $h_j = \mathsf{H}\,(j)$ for different rounds j.

PPP3 assumes that anyone can decrypt consolidated monetary values $c_j^\$$, and therefore, anyone knows $s \overset{\text{def.}}{=} \sum k_i$. If the decryption is not supposed to be public, then attackers can eventually disclose s. Thus, they can decrypt consolidated monetary values $c_j^\$$. Attackers can guess or know

$$c_{j_0}^\$ = \sum_{i=1}^{\tilde{\imath}} m_{i,j_0}$$

for anytime j_0. Afterwards, they can compute

$$g^s = g^{-\tilde{\imath}\cdot h_{j_0}} \cdot \prod_i \mathfrak{M}_{i,j_0} \cdot (1+n)^{-c_{j_0}^\$} \mod n^2,$$

where $c_{j_0}^\$$ is the consolidated monetary value for the round j_0 and $\tilde{\imath}$ is the number of users. After that, they can learn

$$c_j^\$ = \sum_i^{\tilde{\imath}} m_{i,j}$$

for arbitrary round j. If s should be secret, users and their counting agent should use the ADC-Net given by Eq. (6.28) instead of the ADC-Net given by Eq. (6.32).

6.4.7 Performance Analysis

PPP3 gives us consolidated monetary values $c_j^\$$ with verifications. PPP1 gives us consolidated monetary values $c_j^\$$, and PPP2 gives us verifications. Running PPP1 in parallel with PPP2 to give the consolidated monetary values $c_j^\$$ with verifications results in two messages per measurement $m_{i,j}$ instead of one message. Other PPPs also need two messages per measurement $m_{i,j}$, e.g., [5]. PPP3 needs only one.

The size of n is determined by the IFP and the size of the keys k_i is determined by the DLP, cf. Table 6.1 and Fig. 6.6. The operation that requires more processing time is modular exponentiation and its time depends on the exponent size [28], cf. Algorithm 16.

PPP3 given by Eq. (6.32) is faster than the ADC-Net given by Eq. (6.28), because $\mathsf{H}\,(j) \cdot k_i$ has around twice as many bits of $\mathsf{H}\,(j) + k_i$ if $\mathsf{H}\,(j)$ and k_i have approx. the same size.

Comparing the exponent sizes, PPP3 tends to have approx. the same processing time as PPP2. In addition, PPP3 returns consolidated monetary values $c_j^\$$ when PPP2 needs a brute force to recover consolidated monetary values $c_j^\$$. Therefore, PPP3 is the most suitable PPP.

6.5   PPP4 Based on Quantum Mechanics

Einstein et al. [18] drew attention to the fact that quantum mechanics has unusual properties in comparison with classical mechanics. Meanwhile, experiments around the world have confirmed quantum theory. Nowadays, these quantum properties can also be used in cryptography, e.g., a test bed showed that quantum cryptography could protect PMUs [23]. In fact, quantum mechanics can be used to construct quantum computers, which are more powerful than classical computers. For example, the former can solve the IFP and DLP in polynomial time [43], while no such solution based on the latter can be found in the literature. More counter-intuitive, quantum computation enables one to find an element in an unsorted database with complexity $O(\sqrt{n})$ [21], where n is the number of elements. A review of search algorithms, quantum computation, and quantum cryptography may be found in [35, 40], and [11], respectively.

PPP4 is presented—without the monetary value—in the literature [8] along with a PPP based on Quantum Key Distribution (QKD). The literature presents different kinds of QKD used in a smart grid [23, 45]. PPP4 is a preliminary protocol based on quantum mechanics, i.e., it leaves many challenges in engineering [24]. Quantum mechanics can be used for cryptography independent of quantum computers. Currently, quantum cryptography is still not used to protect privacy, but it is already used to provide security [36].

6.5.1   Cryptographic Primitives

The Dirac notation is commonly adopted in quantum mechanics. Hence, a symbol called "ket" is used to denote a vector, i.e., $\vec{\psi}_i \overset{\text{def.}}{=} |\psi_i\rangle \overset{\text{def.}}{=} |i\rangle$. A corresponding symbol called "bra" is used to denote the dual vector, i.e., $\vec{\psi}_i^{\,*} \overset{\text{def.}}{=} \langle\psi_i| \overset{\text{def.}}{=} \langle i|$. Thus, the inner product between two vectors $|i\rangle$ and $|j\rangle$ is called "braket" and denoted as $\langle\psi_i|\psi_j\rangle \overset{\text{def.}}{=} \langle i|j\rangle$. Similarly, the outer product between $|i\rangle$ and $|j\rangle$ is denoted as $|i\rangle\langle j|$, and the tensor product between $|i\rangle$ and $|j\rangle$ is denoted as $|i\rangle \otimes |j\rangle \overset{\text{def.}}{=} |i\rangle|j\rangle \overset{\text{def.}}{=} |i,j\rangle$.

We need to assume four postulates of quantum mechanics.

The first postulate states that a unit vector in a Hilbert space can completely describe the state of any isolated system. Similar to classical bits, the smallest unit of quantum information is a quantum bit called qubit and may be seen as a two-state description. Thus, a qubit is described by a two-dimensional space, i.e., we can write an arbitrary qubit as $|\psi\rangle \overset{\text{def.}}{=} \alpha|0\rangle + \beta|1\rangle$, where $\alpha, \beta \in \mathbb{C}$ and $|\alpha|^2 + |\beta|^2 = 1 = \langle\psi|\psi\rangle$. Similarly, a qudit is a unit vector in a d-dimensional space.

The second postulate states that a unitary matrix describes the evolution of a closed quantum system. Often, the matrices are named either transformations or operators. Therefore, the steps of PPP4 are described by the composition of unitary transformations acting on a vector space.

The third postulate states that the state space of a composite system is the tensor product of its components, e.g., the state of two-qubit composite system $|1\rangle$ and $|0\rangle$ is described by the tensor product $|1\rangle \otimes |0\rangle = |1, 0\rangle$. Surprisingly, not all two-qubit state can be decomposed into the tensor product of two qubits, e.g., the linear combination given by

$$\frac{1}{\sqrt{2}}|1\rangle \otimes |0\rangle + \frac{1}{\sqrt{2}}|0\rangle \otimes |1\rangle = \frac{1}{\sqrt{2}}\left(|1, 0\rangle + |0, 1\rangle\right)$$

cannot be decomposed into the tensor product of two qubits. The impossibility of decomposition of any state is called entanglement, and we say that the state is entangled.

The fourth postulate states that the process of retrieving the information of a state is given by a quantum measurement, and a set $\{M_m\}$ of measurement operators describes quantum measurements, where the index m refers to the possible output from the measurement equipment. Thus, before the quantum measurement, each output is associated with a probability, and after the quantum measurement, we have the state of the system. Therefore, a measurement in quantum mechanics is probabilistic and irreversible.

With these postulates, PPP4 can be constructed. Its first version was published in [8]. Briefly, the counting agent creates an entangled state and sends it to the users who encrypt their measurements $m_{i,j}$ with the entangled state and send results to the counting agent. To decrypt, the counting agent makes a quantum measurement and gets the consolidated monetary value.

6.5.2 Proposed Protocol

In contrast to previous PPPs presented in this chapter, PPP4 does not have a set-up phase and users do not need to have a key. PPP4 assumes that the communication is authenticated, but an attacker might try to read the messages. The counting agent can be curious and try to read individual measurements $m_{i,j}$.

The counting agent starts the protocol with N particles s.t. N is equal to or bigger than the maximum number of users $\tilde{\imath}$ and biggest possible consolidated monetary value. Then, the counting agent prepares an entangled state

$$|E_0\rangle \stackrel{\text{def.}}{=} \frac{1}{\sqrt{N+1}} \sum_{n=0}^{N} \left| \tilde{\imath}\,(N-n), \underbrace{n, n, \ldots, n}_{\tilde{\imath}\ \text{times}} \right\rangle \tag{6.43}$$

and sends the site U_i to the user meter, where the sites are defined by $|\psi_0\rangle_S \otimes |\psi_1\rangle_{U_1} \otimes \cdots \otimes |\psi_{\tilde{\imath}}\rangle_{U_{\tilde{\imath}}} = |\psi_0, \psi_1, \ldots, \psi_{\tilde{\imath}}\rangle$ and the site S belongs to the counting agent. Thus, each user accesses only one site.

After receiving the site, each user i encrypts the measurement $m_{i,j}$ with the phase shifting operation $\exp(\imath\,\hat{N}_{U_i}\delta_i)$, where $\hat{N}_{U_i}|n\rangle_{U_i} = n|n\rangle_{U_i}$,

$$\delta_i \stackrel{\text{def.}}{=} \frac{2\pi\,\text{Value}\,(m_{i,j})}{N+1}, \tag{6.44}$$

and $\imath$ is the imaginary unit. Hereafter, the state of the composite system is changed. After the first user encrypts, we have

$$|E_1\rangle = \frac{1}{\sqrt{N+1}} \sum_{n=0}^{N} \exp(\imath\,n\delta_1)|\,\tilde{\imath}\,(N-n), n, n, \ldots, n\rangle.$$

After the second user, we have

$$|E_2\rangle = \frac{1}{\sqrt{N+1}} \sum_{n=0}^{N} \exp(\imath\,n(\delta_1 + \delta_2))|\,\tilde{\imath}\,(N-n), n, n, \ldots, n\rangle. \tag{6.45}$$

After all $\tilde{\imath}$ users encrypt their measurements $m_{i,j}$, we have

$$|E_{\tilde{\imath}}\rangle = \frac{1}{\sqrt{N+1}} \sum_{n=0}^{N} \exp(\imath\,n\Delta)|\,\tilde{\imath}\,(N-n), n, n, \ldots, n\rangle,$$

where

$$\Delta = \sum_{i=1}^{\tilde{\imath}} \delta_i. \tag{6.46}$$

The consolidated monetary value is given by

$$c_j^{\$} \stackrel{\text{def.}}{=} \sum_{i=1}^{\tilde{\imath}} \text{Value}\,(m_{i,j}), \tag{6.47}$$

Thus, substituting Eqs. (6.44) and (6.47) into Eq. (6.46), we have

$$\Delta = \frac{2\pi c_j^{\$}}{N+1}. \tag{6.48}$$

Before users send their encrypted measurement, the counting agent has access only to the mixed state

$$\mathrm{Tr}_{U_1 \cdots U_{\tilde{\imath}}}(|E_{\tilde{\imath}}\rangle\langle E_{\tilde{\imath}}|) = \frac{1}{N+1} \sum_{n=0}^{N} (|n\rangle\langle n|)_S, \tag{6.49}$$

because only the site S is accessible. Similarly, each user i has access only to the site U_i, and therefore, the mixed state

$$\mathrm{Tr}_{SC_1 \cdots U_{i-1} U_{i+1} \cdots U_{\tilde{\imath}}}(|E_{\tilde{\imath}}\rangle\langle E_{\tilde{\imath}}|) = \frac{1}{N+1} \sum_{n=0}^{N} (|n\rangle\langle n|)_{U_i}. \tag{6.50}$$

After the encryption, users send their site U_i back to the counting agent. Decrypt means to make a quantum measurement with the following states

$$|T_n\rangle = \frac{1}{\sqrt{N+1}} \sum_{k=0}^{N} \exp(\imath k\theta_n)|\tilde{\imath}(N-k), k, k, \ldots, k\rangle, \tag{6.51}$$

where $\theta_n = 2\pi n/(N+1)$. Note that $\langle T_n|T_m\rangle = \delta_{nm}$ for all $n, m \in \{0, .., N\}$, where δ_{nm} is the Dirac delta function. Thus, $\{|T_n\rangle : n = 0, \ldots, N\}$ is an orthonormal basis. Then, the states $|E_i\rangle$ are all eigenvectors of the operator

$$\hat{T} = \sum_{n=0}^{N} n|T_n\rangle\langle T_n|, \tag{6.52}$$

where $|T_n\rangle\langle T_n|$ is the projector onto the eigenspace of $\hat{T}$ with eigenvalue n. Thus, the projective measurement gives us the average value of the measurement that is the expectation value of $\hat{T}$, which can be found by $\langle E_m|\hat{T}|E_m\rangle$. Note that

$$\begin{aligned}
\langle E_{\tilde{\imath}}|T_n\rangle &= \frac{1}{N+1} \sum_{k=0}^{N} \exp(\imath k(\theta_n - \Delta)) \\
&= \frac{1}{N+1} \frac{\exp(\imath \alpha_n(N+1)) - 1}{\exp(\imath \alpha_n) - 1},
\end{aligned} \tag{6.53}$$

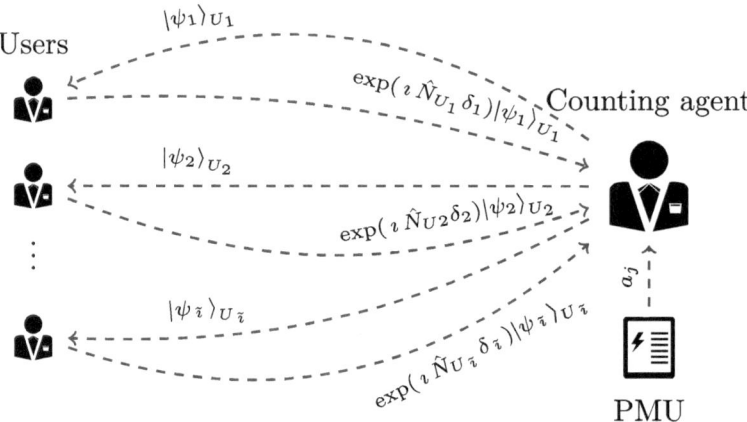

Fig. 6.8 Quantum communication model for PPP4

where $\alpha_n = \theta_n - \Delta = 2\pi(n - c_j^{\$})/(N + 1)$. From Eq. (6.53), we have

$$\langle E_{\tilde{\imath}}|\hat{T}|E_{\tilde{\imath}}\rangle = \sum_{n=0}^{N} n\,\langle E_{\tilde{\imath}}|T_n\rangle\,\langle T_n|E_{\tilde{\imath}}\rangle$$

$$= \frac{1}{(N+1)^2}\sum_{n=0}^{N} n\left(\frac{1 - \cos(\alpha_n(N+1))}{1 - \cos\alpha_n}\right) \tag{6.54}$$

$$= \frac{c_j^{\$}}{\tilde{\imath}}.$$

Therefore, the counting agent decrypts the encrypted consolidated consumption and gets the $c_j^{\$}$ by means of a quantum measurement. Figure 6.8 depicts the communication between the users and their counting agent.

6.5.3   Security Analysis

In contrast to the first three PPPs presented in this chapter, PPP4 is based on postulates of physics, i.e., it is based on the description of physical reality given by quantum mechanics instead of mathematical problems. Equation (6.50) ensures that a user cannot get information from others, and neither can the counting agent due to Eq. (6.49). The counting agent could prepare a non-entangled state and send it to a user. However, the users can check whether their sites are entangled with each other. If a user checks, all users will realize that the entangled sites are being checked.

On the one hand, PPP4 as well as PPP1 can only provide the consolidated monetary value $c_j^\$$. They do not enable verification of the measurements nor bill $b_i^\$$. Thus, a user i can disrupt the communication by sending a huge measurement $m_{i,j}$, and the counting agent cannot detect i. On the other hand, PPP4 is resistant against quantum computers. PPP1 might also be resistant depending on the hash function H. In general, SDC-Nets can provide perfect secrecy under the restriction that the keys be used only once. Protocols based on IFP or DLP are not resistant against quantum computers. There is a research area called post-quantum cryptography whose idea is to develop classical cryptographic algorithms resistant against quantum attackers [2]. Currently, one of its challenges is reducing the processing time and key size [10]. Considering Shor's algorithm [43] and the threatening construction of a quantum computer [1], quantum cryptography and post-quantum cryptography are promising research areas.

6.5.4   Privacy Analysis

To decrypt the consolidated monetary value $c_j^\$$, the counting agent needs to make a quantum measurement of all sites. Quantum measurements of individual sites for the counting agent to obtain an individual measurement $m_{i,j}$ destroy the other sites, because they are entangled. If the counting agent colludes with $\tilde{\imath} - 1$ users and they try to make a quantum measurement in all $\tilde{\imath} - 1$ sites, then the missing site becomes inaccessible. However, the counting agent can collude with $\tilde{\imath} - 1$ users to get the measurements from 1 user. Nevertheless, the collusion with $\tilde{\imath} - 2$ users is not enough. The counting agent can access the consolidated monetary value $c_j^\$$ only once, because the quantum measurements are irreversible. Since they are also probabilistic, quantum measurement without all particles reduce the accuracy of the result. The particles used to create the entangled state return from the users' site to the counting agent's site. Therefore, the counting agent can only access all particles, but if not, only partial information is accessible.

References

1. D. Bacon, D. Leung, Toward a world with quantum computers. Commun. ACM **50**(9), 55–59 (2007). issn:0001-0782. doi:10.1145/1284621.1284648. http://doi.acm.org/10.1145/1284621. 1284648
2. D.J. Bernstein, Introduction to post-quantum cryptography. English, in *Post-Quantum Cryptography*, ed. by D.J. Bernstein, J. Buchmann, E. Dahmen (Springer, Berlin, Heidelberg, 2009), pp. 1–14. isbn:978-3-540-88701-0. doi:10.1007/978-3-540-88702-7_1. http://dx.doi. org/10.1007/978-3-540-88702-7_1
3. R. Böhme et al. On the PET workshop panel 'mix cascades versus peer-to-peer: is one concept superior?' English, in *Privacy Enhancing Technologies*, ed. by D. Martin, A. Serjantov, vol. 3424. Lecture Notes in Computer Science (Springer, Berlin, Heidelberg, 2005), pp. 243–255. isbn:978-3-540-26203-9. doi:10.1007/11423409_16. http://dx.doi.org/10.1007/ 11423409_16

4. D. Boneh, M. Franklin, Efficient generation of shared RSA keys. J. ACM **48**(4), 702–722 (2001). issn:0004- 5411. doi:10.1145/502090.502094. http://doi.acm.org/10.1145/502090. 502094
5. F. Borges, L.A. Martucci, iKUP keeps users' privacy in the Smart Grid, in *2014 IEEE Conference on Communications and Network Security (CNS)* (2014), pp. 310–318. doi:10.1109/CNS.2014.6997499
6. F. Borges, M. Mühlhäuser, EPPP4SMS: efficient privacy-preserving protocol for smart metering systems and its simulation using real-world data. IEEE Trans. Smart Grid **5**(6), 2701–2708 (2014). doi:10.1109/TSG.2014.2336265. http://dx.doi.org/10.1109/TSG.2014.2336265
7. F. Borges, A. Petzoldt, R. Portugal, Small private keys for systems of multivariate quadratic equations using symmetric cryptography, in *XXXIV CNMAC - Congrasso Nacional de Matemática Aplicada e Computacional*. Águas de Lindóia - SP (2012), pp. 1085–1091. http://www.sbmac.org.br/eventos/cnmac/xxxiv_cnmac/pdf/578.pdf
8. F. Borges et al., A privacy-enhancing protocol that provides in-network data aggregation and verifiable smart meter billing, in *2014 IEEE Symposium on Computers and Communication (ISCC)* (2014), pp. 1–6. doi:10.1109/ISCC.2014.6912612
9. F. Borges, J. Buchmann, M. Mühlhäuser, Introducing asymmetric DC-Nets, in *2014 IEEE Conference on Communications and Network Security (CNS)* (2014), pp. 508–509. doi:10.1109/CNS.2014.6997528
10. F. Borges, R.A.M. Santos, F.L. Marquezino, Preserving privacy in a smart grid scenario using quantum mechanics. Secur. Commun. Netw., n/a (2014). issn:1939-0122. doi:10.1002/sec.1152. http://dx.doi.org/10.1002/sec.1152
11. D. Bruss et al., Quantum cryptography: a survey. ACM Comput. Surv. **39**(2) (2007). issn:0360-0300. doi:10.1145/1242471.1242474. http://doi.acm.org/10.1145/1242471.1242474.
12. J. Camenisch, A. Lysyanskaya, M. Meyerovich, Endorsed e-cash, in *IEEE Symposium on Security and Privacy, 2007. SP '07* (2007), pp. 101–115. doi:10.1109/SP.2007.15
13. D. Chaum, The dining cryptographers problem: unconditional sender and recipient untraceability. J. Cryptol. **1**(1), 65–75 (1988). issn:0933-2790. http://dl.acm.org/citation.cfm?id=54235.54239.
14. H. Cohen et al., *Handbook of Elliptic and Hyperelliptic Curve Cryptography*, 2nd edn. (Chapman & Hall/CRC, London/Boca Raton, 2012). isbn:1439840008, 9781439840009
15. R. Cramer, R. Gennaro, B. Schoenmakers, A secure and optimally efficient multi-authority election scheme, in *Proceedings of the 16th Annual International Conference on Theory and Application of Cryptographic Techniques*. EUROCRYPT'97 (Springer, Konstanz, 1997), pp. 103–118. isbn:3-540-62975-0. http://dl.acm.org/citation.cfm?id=1754542.1754554
16. R. Cramer, I. Damgård, J.B. Nielsen, Multiparty computation from threshold homomorphic encryption, in *Proceedings of the International Conference on the Theory and Application of Cryptographic Techniques: Advances in Cryptology*. EUROCRYPT '01 (Springer, London, 2001), pp. 280–299. isbn:3-540-42070-3. http://dl.acm.org/citation.cfm?id=647086.715687
17. T. Dimitriou, G. Karame, I. Christou, SuperTrust a secure and efficient framework for handling trust in super peer networks, in *Proceedings of the 9th International Conference on Distributed Computing and Networking*. ICDCN'08 (Springer, Kolkata, 2008), pp. 350–362. isbn:3-540-77443-2, 978-3-540-77443-3. http://dl.acm.org/citation.cfm?id=1785854.1785901
18. A. Einstein, B. Podolsky, N. Rosen, Can quantum-mechanical description of physical reality be considered complete? Phys. Rev. **47**, 777–780 (1935). doi:10.1103/PhysRev.47.777. http://link.aps.org/doi/10.1103/PhysRev.47.777
19. P. Erdös, C. Pomerance, E. Schmutz, Carmichael's lambda function. Acta Arith **58**(4), 363–385 (1991)
20. C. Gentry, A fully homomorphic encryption scheme. Ph.D. thesis. Stanford University (2009). crypto.stanford.edu/craig
21. L.K. Grover, A fast quantum mechanical algorithm for database search, in *Proceedings of the Twenty-eighth Annual ACM Symposium on Theory of Computing*. STOC '96 (ACM, Philadelphia, PA, 1996), pp. 212–219. isbn:0-89791-785-5. doi:10.1145/237814.237866. http://doi.acm.org/10.1145/237814.237866

22. D. Hankerson, A.J. Menezes, S. Vanstone, *Guide to Elliptic Curve Cryptography* (Springer, New York, Secaucus, NJ, 2003). isbn:038795273X
23. R.J. Hughes et al., Network-centric quantum communications with application to critical infrastructure protection. ArXiv e-prints (2013). arXiv:1305.0305 [quant-ph]
24. S. Imre, Quantum communications: explained for communication engineers. IEEE Commun. Mag. **51**(8), 28–35 (2013). issn:0163-6804. doi:10.1109/MCOM.2013.6576335
25. F. Kerschbaum, A verifiable, centralized, coercion-free reputation system, in *Proceedings of the 8th ACM Workshop on Privacy in the Electronic Society*. WPES '09 (ACM, Chicago, IL, 2009), pp. 61–70. isbn:978-1-60558-783-7. doi:10.1145/1655188.1655197. http://doi.acm. org/10.1145/1655188.1655197
26. R.E. Klima, *Applications of Abstract Algebra with Maple and MATLAB (Discrete Mathematics and Its Applications)* (Chapman & Hall/CRC, London/Boca Raton, 2007). isbn:1420011197
27. N. Koblitz, Elliptic curve cryptosystems. Math. Comput. **48**(177), 203–209 (1987). issn:00255718
28. P. Lara et al., Parallel modular exponentiation using load balancing without precomputation. J. Comput. Syst. Sci. **78**(2), 575–582 (2012). issn:0022-0000. doi:10.1016/j.jcss.2011.07.002. http://dx.doi.org/10.1016/j.jcss.2011.07.002
29. Q. Li, G. Cao, Efficient privacy-preserving stream aggregation in mobile sensing with low aggregation error. English, in *Privacy Enhancing Technologies* ed. by E. De Cristofaro, M. Wright, vol. 7981. Lecture Notes in Computer Science (Springer, Berlin, Heidelberg, 2013), pp. 60–81. isbn:978-3-642-39076-0. doi:10.1007/978-3-642-39077-7_4. http://dx.doi.org/10. 1007/978-3-642-39077-7_4
30. F. Li, B. Luo, Preserving data integrity for smart grid data aggregation, in *2012 IEEE Third International Conference on Smart Grid Communications (SmartGridComm)* (2012), pp. 366–371. doi:10.1109/SmartGridComm.2012.6486011
31. F. Li, B. Luo, P. Liu, Secure information aggregation for smart grids using homomorphic encryption, in *2010 First IEEE International Conference on Smart Grid Communications (SmartGridComm)* (2010), pp. 327–332. doi:10.1109/SMARTGRID.2010.5622064
32. A. Menezes, S. Vanstone, T. Okamoto, Reducing elliptic curve logarithms to logarithms in a finite field, in *Proceedings of the Twenty-Third Annual ACM Symposium on Theory of Computing*. STOC '91 (ACM, New Orleans, LA, 1991), pp. 80–89. isbn:0-89791-397-3. doi:10.1145/103418.103434
33. V.S. Miller, Use of elliptic curves in cryptography, in *Advances in Cryptology—CRYPTO 85*. Lecture Notes in Computer Science, vol. 218 (Springer, New York, Santa Barbara, CA, 1986), pp. 417–426. isbn:0-387-16463-4
34. A. Molina-Markham et al., Designing privacy-preserving smart meters with low-cost microcontrollers, in *Financial Cryptography*, vol. 7397. Lecture Notes in Computer Science (Springer, Berlin, 2012), pp. 239–253. isbn:978-3-642-32945-6
35. M.A. Nielsen, I.L. Chuang, *Quantum Computation and Quantum Information*. Cambridge Series on Information and the Natural Sciences (Cambridge University Press, Cambridge, 2000). isbn:9780521635035. http://books.google.de/books?id=65FqEKQOfP8C
36. M. Niemiec, A.R. Pach, Management of security in quantum cryptography. IEEE Commun. Mag. **51**(8), 36–41 (2013). issn:0163-6804. doi:10.1109/MCOM.2013.6576336
37. P. Paillier, Public-key cryptosystems based on composite degree residuosity classes, in *Advances in Cryptology EUROCRYPT 1999*, vol. 1592. Lecture Notes in Computer Science (Springer, Berlin, 1999), pp. 223–238. isbn:978-3-540-65889-4
38. T.P. Pedersen, Non-interactive and information-theoretic secure verifiable secret sharing, in *Proceedings of the 11th Annual International Cryptology Conference on Advances in Cryptology*. CRYPTO '91 (Springer, London, 1992), pp. 129–140. isbn:3-540-55188-3. http:// dl.acm.org/citation.cfm?id=646756.705507
39. S. Peter, D. Westhoff, C. Castelluccia, A survey on the encryption of convergecast traffic with in-network processing. IEEE Trans. Dependable Secure Comput. **7**(1), 20–34 (2010). issn:1545-5971. doi:10.1109/TDSC.2008.23

40. R. Portugal, *Quantum Walks and Search Algorithms* (Springer, Berlin, 2013). isbn:1461463351, 9781461463351
41. S. Ruj, A. Nayak, A decentralized security framework for data aggregation and access control in smart grids. IEEE Trans. Smart Grid **4**(1), 196–205 (2013). issn:1949-3053. doi:10.1109/TSG.2012.2224389
42. R. Schoof, Counting points on elliptic curves over finite fields. English. Journal de théorie des nombres de Bordeaux **7**(1), 219–254 (1995). http://eudml.org/doc/247664
43. P.W. Shor, Polynomial-time algorithms for prime factorization and discrete logarithms on a quantum computer. SIAM J. Comput. **26**(5), 1484–1509 (1997). issn:0097-5397. doi:10.1137/S0097539795293172.
44. N.P. Smart, The discrete logarithm problem on elliptic curves of trace one. J. Cryptol. **12**(3), 193–196 (1999)
45. S.M. Zhang, X.Y. Liu, B.Y. Wang, An applied research of improved BB84 protocol in electric power secondary system communication. English, in *Advances in Electronic Engineering, Communication and Management Vol. 1*, ed. by D. Jin, S. Lin, vol. 139. Lecture Notes in Electrical Engineering (Springer, Berlin, Heidelberg, 2012), pp. 545–550. isbn:978-3-642-27286-8. doi:10.1007/978-3-642-27287-5_87. http://dx.doi.org/10.1007/978-3-642-27287-5_87
46. P. Zheng, J. Huang, An efficient image homomorphic encryption scheme with small ciphertext expansion, in *Proceedings of the 21st ACM International Conference on Multimedia* MM '13 (ACM, Barcelona, 2013), pp. 803–812. isbn:978-1-4503-2404-5. doi:10.1145/2502081.2502105. http://doi.acm.org/10.1145/2502081.2502105

Chapter 7
Analytical Comparison

Abstract This chapter presents an analytical comparison between Privacy-Preserving Protocols (PPPs) for smart metering systems. Specifically, it compares the PPPs described in Chaps. 3 and 6 with each other. In particular, this analytical comparison for PPPs focuses on security, privacy, requirements, verification property, and performance.

Keywords Security • Privacy • Requirements • Verification • Performance • Complexity

To provide a more general comparison, the PPPs in Chap. 3 represent a class of protocols. Specifically, Algorithm 1 [19] is an additive homomorphic encryption primitive (AHEP) used in many PPPs for smart grids, e.g., [17, 21]. Instead of comparing protocols that use Algorithm 1, we can have a class of "AHEP" to compare. Another class is "homomorphic signature (HS)," which also has been used in PPPs for smart grids, e.g., [16, 22]. Symmetric DC-Nets (SDC-Nets) can be connected in many ways, e.g., star [5] and randomly connected [1]. However, similar to PPP1, they need a specific trust model with more assumptions than a fully connected SDC-Net [10, 14]. To keep the comparison more general, this chapter addresses the class "SDC-Net" as a fully connected SDC-Net, cf. Algorithm 2. The class "matching" represents PPPs with a homomorphic commitment scheme with a matching AHEP, e.g., [4]. One can use multiple protocols together to compose a protocol that can fulfill the requirements in Sect. 4.2, for instance, PPP1 running along with PPP2. Another example is found in [5]. However, the performance of such multi-protocols is dependent on the sum of the cost of their protocols, and security and privacy are based on the weakest protocol. In contrast to the classes, Chap. 6 presents the PPPs one after another. Particularly, PPP3 represents Asymmetric DC-Nets (ADC-Nets). A comparison between individual protocols can be found in [4, 6, 7].

Sections 7.1 and 7.2 present a discussion about security and privacy, respectively. Section 7.3 distinguishes the PPPs by requirements as described in Sect. 4.2, namely recoverability of bill $b_i^\$$ and of consolidated monetary value, verification, and computational efficiency. The two last requirements have several points to be

© Springer International Publishing Switzerland 2017   101
F. Borges de Oliveira, *On Privacy-Preserving Protocols for Smart Metering Systems*,
DOI 10.1007/978-3-319-40718-0_7

analyzed. For this reason, Sects. 7.4 and 7.5 present a discussion about verification and performance, respectively. Section 7.6 finalizes this chapter with a comparison between SDC-Nets, AHEPs, and ADC-Nets.

7.1 Security

Table 7.1 shows to what extent the protocols have the following properties: freedom of trusted third party (TTP), anti-collusion, and fault tolerance. Considering that only the counting agent has the private key, aggregators are semi-trusted parties. Thus, TTP is any organization used in the protocol to do more than aggregation. For example, a TTP can set up the keys. Anti-collusion ensures that the counting agent cannot collude with a TTP or with the aggregator. Protocols using in-network aggregation have a virtual aggregator; cf. Sect. 3.2.1. However, parts of such a virtual aggregator can collude with the counting agent. Access control can be used to avoid collusion [21], but this is not an intrinsic property of AHEP neither PPP1. Moreover, one who grants access might also collude. Similarly, fault tolerance with respect to the communication can be achieved with a TTP [12], but again it is an intrinsic property of neither AHEP nor PPP1. Nevertheless, protocols are fault tolerant when they allow the counting agent to identify the failures in the communication and to restart the protocols without the meters with failures in the communication, e.g., a commitment function with digital signature. Multi-protocols may be fault tolerant, depending on their parts; for instance, if a multi-protocol uses an SDC-Net and a commitment function with digital signature, then it can be fault tolerant. Contrarily, if one part is not fault tolerant, the whole multi-protocol cannot be, for instance, [5]. Fault tolerance with respect to processing and storage is not in the scope of PPPs.

In general, the security of PPPs depends on mathematical problems, e.g., IFP, DLP, and ECDLP. Differently, commitments can be unconditionally secure [20] depending on the group used. Similarly, SDC-Net can be unconditionally secure if

Table 7.1 Comparison of secure

Protocol	TTP free	Anti-collusion	Fault Tolerance	Problem
AHEP	Yes	No	No	IFP,DLP, etc.
SDC-Net	Yes	Yes	Yes	H
HS	Yes	No	No	ECDLP
Commitments	Yes	Yes	Yes	IFP,DLP, etc.
Matching	Yes	No	No	IFP,DLP, etc.
PPP1	Yes	No	No	H
PPP2	Yes	Yes	Yes	H and ECDLP
PPP3	Yes	Yes	Yes	H, IFP, and DLP
PPP4	Yes	Yes	No	quantum cryptography

the keys are used only once [9]. In practice, commitment schemes depend on digital signatures, and SDC-Nets depend on a secure hash function s.t. it behaves as a one-way function and has collision resistance. The security of the four protocols presented in Chap. 6 depends on a hash function H, a hash function H_Ω with ECDLP, a hash function H with DLP and IFP, and entanglement, respectively. Instead of breaking the protocols, an attacker might get some information depending on the communication network as described in Sect. 2.2. Sections 6.2.1, 6.3.3, 6.4.5, and 6.5.3 present a discussion about security of PPP1 to PPP4, respectively.

In summary, only SDC-Net, Commitments, PPP2, and PPP3 are simultaneously free of TTP, anti-collusion, and fault tolerant. The differences between them can be seen in the next sections.

7.2 Privacy

The second column of Table 7.2 indicates the protocols that need some sort of aggregator other than the counting agent. A virtual aggregator can be done with in-network aggregation, but it is also an aggregator; cf. Sect. 3.2.1. The third column shows the number of users required in collusion to cause a leak of privacy. The fourth column indicates if the counting agent needs to collude. The last column shows if the counting agent is able to decrypt individual measurements $m_{i,j}$.

Although commitments might be unconditionally secure, $\tilde{\imath} - 1$ users can collude to disclose information from one user. PPP4 enables the counting agent to decrypt individual measurements $m_{i,j}$. However, the counting agent loses the consolidated monetary value $c_j^\$$. The number $\tilde{\imath} - 1$ for collusion is the minimal number of users necessary to disclose the measurements of one user. This is a threshold for all PPPs including SDC-Nets that can be considered unconditionally secure when the keys are used only once and there is no collusion. The counting agent can have a key in schemes based on SDC-Nets and ADC-Nets. In this scenario, the counting agent should also collude, but this is not the setup used.

Table 7.2 Comparison of privacy

Protocol	Aggregator	Collusion	Counting agent	$m_{i,j}$
AHEP	Yes	1	Yes	Yes
SDC-Net	No	$\tilde{\imath} - 1$	No	No
HS	Yes	1	Yes	Yes
Commitments	No	$\tilde{\imath} - 1$	Yes	No
Matching	Yes	1	Yes	Yes
PPP1	Yes	1	Yes	Yes
PPP2	No	$\tilde{\imath} - 1$	Yes	No
PPP3	No	$\tilde{\imath} - 1$	Yes	No
PPP4	No	$\tilde{\imath} - 1$	No	Yes

In summary, only SDC-Net, Commitments, PPP2, and PPP3 are the better protocols for privacy. They do not need an external aggregator. In addition, they only reveal information from a user if all others collude, and the counting agent needs to collude. Moreover, the counting agent cannot decrypt individual measurements $m_{i,j}$.

7.3 Requirements

This section shows the requirements fulfilled by PPP. They are described in Sect. 4.2, namely: recoverability of bill $b_i^\$$, recoverability of consolidated consumption, verification, and computational efficiency. The PPPs should fulfill each requirement in polynomial time. Otherwise, the requirement is not considered fulfilled. For example, all protocols in Chap. 6 can retrieve consolidated monetary values $c_j^\$$. However, PPP2 needs an exponential complexity time to find the correct $c_j^\$$. Thus, if a user tries to disrupt the communication sending a huge measurement $m_{i,j}$, the consolidated monetary value is also huge and the counting agent needs to solve the ECDLP to find $c_j^\$$. Therefore, PPP2 does not provide consolidated monetary values $c_j^\$$ in polynomial time. Note that users would not become attackers because they can be discovered with PPP2's verification.

The second column of Table 7.3 indicates the protocols that used aggregated measurements a_j to validate consolidated monetary values. Although few of them had used a_j [4], all of them can use it. The third column indicates the protocols that are considered $b_i^\$$. In particular, Jawurek et al. [13] presented a commitment scheme in which the measurements $m_{i,j}$ can stay in the meters, which only send commitments to the supplier. Afterwards, Molina-Markham et al. [18] presented the performance of different commitments in low-cost microcontrollers. Despite $b_i^\$$ already being required in a non-smart grid, the majority of the published papers have only considered consolidated consumptions c_j. For example, PPPs [17, 21] based on Paillier [19], HS [16, 22], SDC-Net as protocols presented in [1, 10, 14], and

Table 7.3 Comparison of requirements between PPP

Protocol	a_j	$b_i^\$$	$c_j^\$$	Verification	Efficiency
AHEP	No	No	Yes	No	No
SDC-Net	No	No	Yes	No	No
HS	No	No	No	Yes	Yes
Commitments	No	Yes	No	Yes	Yes
Matching	Yes	Yes	Yes	yes	No
PPP1	Yes	No	Yes	No	Yes
PPP2	Yes	Yes	No	Yes	Yes
PPP3	Yes	Yes	Yes	Yes	Yes
PPP4	Yes	No	Yes	No	NA

Table 7.4 Comparison of verification capabilities

Protocol	$b_i^\$$	c_j	Transmission error	Deceptive users
AHEP	No	No	No	No
SDC-Net	No	No	Yes	No
HS	No	Yes	Yes	Yes
Commitments	Yes	No	Yes	No
Matching	Yes	Yes	Yes	Yes
PPP1	No	No	No	No
PPP2	Yes	Yes	Yes	Yes
PPP3	Yes	Yes	Yes	Yes
PPP4	No	No	No	No

even protocols based on quantum mechanics [8]. The fourth column indicates the protocols that have provided $c_j^\$$ in polynomial time. The fifth indicates protocols that have considered verification. Section 7.4 presents a discussion about verification. The last column provides an overview of the efficiency, which has a description in Sect. 7.5. The efficiency of PPP4 depends on quantum mechanics. For this reason, its spot is labeled not applicable (NA).

In summary, only PPP3 has addressed all the requirements.

7.4 Verification Property

Users and their counting agent should be able to verify whether the values of bill $b_i^\$$ and consolidated consumption c_j are correct. In addition, the counting agent should be able to verify whether there are errors in message transmissions, energy losses, or frauds, and if some users are deceptive. In practice, deceptive users may represent failures in measurement, processing, or communication. Moreover, PPPs should detect the meters with failures. Table 7.4 shows these verification capabilities. HS might detect deceptive users, if the counting agent stores all measurements and searches on them, resulting in invasion of privacy. In addition, it is possible to enable billing verification under the same circumstances, but it has not been considered in the literature. One might say that protocols based on Paillier can have all properties in Table 7.4. However, it can be achieved only with a TTP, similar to PPP1.

To perform the verification, the counting agent needs to store only one product of encrypted measurements $\mathfrak{M}_{i,j}$ per verifiable value [4], e.g., one product per bill $b_i^\$$.

In summary, only matching, PPP2, and PPP3 have full capabilities of verification.

7.5 Performance

Asymptotic complexity analysis gives an overview of the performance. However, two algorithms can have the same complexity, but one can be much more efficient than another can. For example, the Paillier scheme works on a group $\mathbb{Z}_n$, where

Table 7.5 Comparison of processing time

Protocol	Encryption	Aggregation	Decryption
AHEP	$O(\log(n))$	$O(\tilde{\imath})$	$O(\log(n))$
SDC-Net	$O(\tilde{\imath})$	NA	$O(\tilde{\imath})$
HS	$O(\log(k))$	$O(\tilde{\imath})$	$O(k)$
Commitments	$O(\log(k))$	$O(\tilde{\jmath})$	$O(k)$
Matching	$O(\log(n))$	$O(\tilde{\imath})$	$O(\log(n))$
PPP1	$O(1)$	$O(\tilde{\imath})$	$O(\tilde{\imath})$
PPP2	$O(\log(k))$	$O(\tilde{\imath})$	$O(k)$
PPP3	$O(\log(k))$	$O(\tilde{\imath})$	$O(\log(k))$

n is the product of two safe primes, and a matching [4] works on a group $\mathbb{Z}_{4n+1}$. Thus, the Paillier scheme is more efficient, but both have the same time complexity $O(\log(n))$, which is determined by the modular exponentiation [15]. Similarly, SDC-Nets have time complexity $O(\tilde{\imath})$ for encryption and decryption, but the former needs to compute $\tilde{\imath}$ hash functions and the latter needs to compute $\tilde{\imath}$ additions. The aggregation for SDC-Nets happens together with the decryption and the processing cannot be split. Thus, Table 7.5 presents this result as NA. Besides SDC-Nets, other protocols have time complexity $O(\tilde{\imath})$ for the aggregation. However, PPP1 needs $\tilde{\imath}$ additions, while PPP2 and HS need $\tilde{\imath}$ operations over elliptic curves. Normally, others need $\tilde{\imath}$ modular multiplications over integers. Thus, the processing times are different although the complexities are equal. Table 7.5 shows the time complexity for encryption, aggregation, and description for respective protocols, where $\tilde{\imath}$ is the number of users, $\tilde{\jmath}$ is the number of rounds, n has size of IFP, and k has size of DLP; cf. Table 6.1 and Fig. 6.6. PPP4 does not appear because its nature is different from the others. The encryption is done with a unitary transformation and the decryption with a quantum measurement. Although commercial products have used entangled states [8, 11], Almeida et al. [2] presented a time limitation for keeping the states entangled, and therefore, a problem for scalability.

There are two more issues for complexity, namely communication and storage. Table 7.6 shows the complexity to set up the protocols, the number of keys per users to protect the privacy, and the number of messages per measurement. In addition, it shows whether users send messages directly to the counting agent. The set-up phase of all protocols can be done rapidly with a TTP, but TTP is a single point of failure. Thus, it should be used only when strictly necessary. In the ideal situation, each user has only one key to protect the privacy and sends only one message per measurement directly to their counting agent who can detect failures in the communication. Commitment schemes do not require keys, thus, this is NA. PPP4 does not require a key stored, but the counting agent should send the particles to each user i for each round j. Matching requires two messages per measurement $m_{i,j}$, and multi-protocols require at least two. Using in-network aggregation, the counting agent might receive only one message per round j, instead of the number of users $\tilde{\imath}$. However, this technique obstructs the counting agent to detect failures in the communication. The overload of the counting agent can be minimized using lossless aggregation [3].

Table 7.6 Comparison of communication

Protocol	Setup	Key	Message	Direct
AHEP	$O(\tilde{\imath})$	1	1	No
SDC-Net	$O(\tilde{\imath}^2)$	$O(\tilde{\imath})$	1	Yes
HS	$O(\tilde{\imath})$	1	1	No
Commitments	$O(\tilde{\imath})$	NA	1	Yes
Matching	$O(\tilde{\imath})$	1	2	No
PPP1	$O(\tilde{\imath})$	1	1	No
PPP2	$O(\tilde{\imath}^2)$	1	1	Yes
PPP3	$O(\tilde{\imath}^2)$	1	1	Yes

In summary, Table 7.5 shows that PPP1 is the most efficient to encrypt, while SDC-Nets are the least efficient. However, SDC-Nets are faster than PPP1 in decryption. SDC-Nets does not need aggregation, the others are equivalent to each other. Excluding PPP1, the best processing time for encryption and decryption is $O(\log(k))$. Table 7.6 shows that SDC-Net, PPP3, and ADC-Net have the heaviest setup, but it is the most secure setup. PPP3 and ADC-Net might use PPP1 in its set-up phase to reduce the communication cost and to have setup $O(\tilde{\imath})$. Besides commitment schemes that do not have cryptographic key, the best protocols are PPP2 and PPP3.

7.6 Summary

Although the literature contains many PPPs, there are few primitives to protect privacy. HS and commitments do not decrypt in polynomial time. In addition, considering matching as AHEPs with a commitment scheme, then we have only SDC-Nets, AHEPs, and ADC-Nets to encrypt and decrypt the measurements. Note that PPP1 is a star SDC-Nets and PPP2 has no decryption as well as other commitment schemes, i.e., the consolidated monetary value $c_j^\$$ cannot be recovered in polynomial time. Table 7.7 summarizes the comparison between SDC-Net, AHEP, and ADC-Net.

ADC-Nets have all the benefits of SDC-Nets and AHEPs. The three primitives enable users to send the minimum number of messages per measurement $m_{i,j}$, users and their counting agent can use permanent keys, and the security is based on a trapdoor. Note that a hash function H should be irreversible and can be considered as a one-way function. Since AHEP and ADC-Net are asymmetrical, they are also based on a cryptographic trapdoor. Only SDC-Nets and ADC-Nets avoid collusion, because $O(\tilde{\imath})$ users should collude to leak information from only one user. Not all users need to be in the aggregation process, i.e., only the set of trusted users in an SDC-Net or an ADC-Net. AHEPs can decrypt individual measurements, thus it does not have the concept of trusted users. They and their counting agent can set up an SDC-Net or an ADC-Net for users to send messages directly to their

Table 7.7 Comparison between SDC-Nets, AHEPs, and ADC-Nets

Properties	SDC-Net	AHEP	ADC-Net
Collusion of $O(\tilde{\iota})$	✓		✓
Set of trusted users	✓		✓
Messages to the counting agent	✓		✓
Minimum number of messages	✓	✓	✓
Scalable		✓	✓
Permanent keys	✓	✓	✓
Based on trapdoors	✓	✓	✓
Keys stored per user	$2(\tilde{\iota}-1)$	1	1
Total of keys	$O(\tilde{\iota}^2)$	2	$O(\tilde{\iota})$
Polynomial time	✓	✓	✓
One cannot disrupt			✓
Verification as commitment			✓

counting agent. Thus, users can sign their messages, and the counting agent can detect failures in the communication. Since the complexity of the algorithms for AHEPs and ADC-Nets has polynomial time over the key sizes, they are scalable. In contrast, each user using a fully connected SDC-Net should iterate over all users, i.e., $\tilde{\iota}$. Thus, SDC-Nets are not scalable in relation to the number of users $\tilde{\iota}$. Note that the counting agent has the same problem using PPP1. Nevertheless, PPP1 is $\tilde{\iota}$ times faster than a fully connected SDC-Net—for instance, LOP—; cf. Sects. 3.2.2 and 6.2. Only AHEPs and SDC-Nets require only one key per user to protect privacy, i.e., without considering the keys to sign the messages. An AHEP has only one public–private key pair. Nevertheless, $\tilde{\iota}$ is the minimum number of keys to avoid decryption of individual measurements. In addition, only ADC-Nets avoid disruption of the communication and have verification like commitment. The former can prevent malicious users from disrupting the communication by injecting huge measurements, because they can be discovered. The latter enable users and their counting agent to verify values—for instance, consolidated consumption c_j and bill $b_i^\$$—similar to commitment schemes.

The benefit of AHEPs is the key distribution, i.e., one public–private key pair. Despite that, SDC-Nets and ADC-Nets can replace AHEPs in PPPs. PPP1 is an example of an SDC-Net that can behave as an AHEP. Moreover, each AHEP is a particular case of an ADC-Net, i.e., ADC-Nets are generalizations of AHEPs, cf. Sect. 6.4.1.

References

1. G. Ács, C. Castelluccia, I have a DREAM! (DiffeRentially privatE smArt Metering), in *Information Hiding: 13th International Conference (IH 2011), Prague, Czech Republic, May 18–20*, ed. by T. Filler et al. (Springer, Berlin/Heidelberg, 2011), pp. 118–132; Revised selected papers, isbn:978-3-642-24178-9, doi:10.1007/978-3-642-24178-9_9, url:http://dx.doi.org/10.1007/978-3-642-24178-9_9

2. M.P. Almeida et al., Environment-induced sudden death of entanglement. Science **316**(5824), 579–582 (2007). doi:10.1126/science.1139892, url:http://dx.doi.org/10.1126/science.1139892

3. A. Bartoli et al., Secure lossless aggregation over fading and shadowing channels for smart grid M2M networks. IEEE Trans. Smart Grid **2**(4), 844–864 (2011). issn: 1949-3053, doi:10.1109/TSG.2011.2162431

4. F. Borges, L.A. Martucci, iKUP keeps users' privacy in the smart grid, in *2014 IEEE Conference on Communications and Network Security (CNS)* (2014), pp. 310–318. doi:10.1109/CNS.2014.6997499

5. F. Borges, M. Mühlhäuser, EPPP4SMS: efficient privacy-preserving protocol for smart metering systems and its simulation using real-world data. IEEE Trans. Smart Grid **5**(6), 2701–2708 (2014). doi:10.1109/TSG.2014.2336265, url:http://dx.doi.org/10.1109/TSG.2014.2336265

6. F. Borges, R.A.M. Santos, F.L. Marquezino, Preserving privacy in a smart grid scenario using quantum mechanics. Security Commun. Netw. (2014); n/a–n/a, issn:1939-0122, doi:10.1002/sec.1152, url:http://dx.doi.org/10.1002/sec.1152

7. F. Borges et al., A privacy-enhancing protocol that provides in-network data aggregation and verifiable smart meter billing, in *2014 IEEE Symposium on Computers and Communication (ISCC)* (2014), pp. 1–6; doi:10.1109/ISCC.2014.6912612

8. F. Borges, F. Volk, M. Mühlhäuser, Efficient, verifiable, secure, and privacy-friendly computations for the smart grid, in *Innovative Smart Grid Technologies Conference (ISGT)* (IEEE Power Energy Society, 2015), pp. 1–5; doi:10.1109/ISGT.2015.7131862

9. D. Chaum, The dining cryptographers problem: unconditional sender and recipient untraceability. J. Cryptol. **1**(1), 65–75 (1988); issn:0933-2790, url:http://dl.acm.org/citation.cfm?id=54235.54239

10. Z. Erkin, G. Tsudik, Private computation of spatial and temporal power consumption with smart meters, in: *Applied Cryptography and Network Security*, vol. 7341, ed. by F. Bao, P. Samarati, J. Zhou. Lecture Notes in Computer Science (Springer, 2012), pp. 561–577; isbn:978-3-642-31283-0

11. S. Imre, Quantum communications: explained for communication engineers. IEEE Commun. Mag. **51**(8), 28–35 (2013); issn:0163-6804. doi:10.1109/MCOM.2013.6576335

12. M. Jawurek, F. Kerschbaum, Fault-tolerant privacy-preserving statistics, in *Privacy Enhancing Technologies*, vol. 7384, ed. by S. Fischer-Hübner, M. Wright. Lecture Notes in Computer Science (Springer, Berlin/Heidelberg, 2012), pp. 221–238; isbn:978-3-642-31679-1, doi:10.1007/978-3-642-31680-7_12, url:http://dx.doi.org/10.1007/978-3-642-31680-7$\delimiter"026E30F$_12

13. M. Jawurek, M. Johns, F. Kerschbaum, Plug-In Privacy for Smart Metering Billing, in *Proceedings of the Privacy Enhancing Technologies: 11th International Symposium, Waterloo, ON, Canada, July 27–29, 2011*, ed. by S. Fischer-Hübner, N. Hopper (Springer, Berlin/Heidelberg, 2011), pp. 192–210; isbn:978-3-642-22263-4, doi:10.1007/978-3-642-22263-4_11, url:http://dx.doi.org/10.1007/978-3-642-22263-4_11

14. K. Kursawe, G. Danezis, M. Kohlweiss, Privacy-friendly aggregation for the smart-grid, in *Proceedings of the Privacy Enhancing Technologies: 11th International Symposium, Waterloo, ON, Canada, July 27–29, 2011*, ed. by S. Fischer-Hübner, N. Hopper (Springer, Berlin/Heidelberg, 2011), pp. 175–191; isbn:978-3-642-22263-4, doi:10.1007/978-3-642-22263-4_10, url:http://dx.doi.org/10.1007/978-3-642-22263-4_10

15. P. Lara et al., Parallel modular exponentiation using load balancing without precomputation. J. Comput. Syst. Sci. **78**(2), 575–582 (2012); issn:0022-0000, doi:10.1016/j.jcss.2011.07.002, url:http://dx.doi.org/10.1016/j.jcss.2011.07.002

16. F. Li, B. Luo, Preserving data integrity for smart grid data aggregation, in *2012 IEEE Third International Conference on Smart Grid Communications (SmartGridComm)* (2012), pp. 366–371; doi:10.1109/SmartGridComm.2012.6486011

17. F. Li, B. Luo, P. Liu, Secure information aggregation for smart grids using homomorphic encryption, in *2010 First IEEE International Conference on Smart Grid Communications (SmartGridComm)* (2010), pp. 327–332; doi:10.1109/SMARTGRID.2010.5622064

18. A. Molina-Markham et al., Designing privacy-preserving smart meters with low-cost microcontrollers, in *Financial Cryptography*, vol. 7397. Lecture Notes in Computer Science (Springer, Berlin, 2012), pp. 239–253; isbn:978-3-642-32945-6
19. P. Paillier, Public-key cryptosystems based on composite degree residuosity classes, in *Advances in Cryptology - EUROCRYPT 1999*, vol. 1592. Lecture Notes in Computer Science (Springer, Berlin, 1999), pp. 223–238; isbn:978-3-540-65889-4
20. T.P. Pedersen, Non-interactive and information-theoretic secure verifiable secret sharing, in *Proceedings of the 11th Annual International Cryptology Conference on Advances in Cryptology* (Springer, London, 1992), pp. 129–140; isbn:3-540-55188-3, url: http://dl.acm.org/citation.cfm?id=646756.705507
21. S. Ruj, A. Nayak, A decentralized security framework for data aggregation and access control in smart grids. IEEE Trans. Smart Grid **4**(1), 196–205 (2013); issn:1949-3053, doi:10.1109/TSG.2012.2224389
22. L. Yang, F. Li, Detecting false data injection in smart grid in-network aggregation, in *2013 IEEE International Conference on Smart Grid Communications (SmartGridComm)* (2013), pp. 408–413; doi:10.1109/SmartGridComm.2013.6687992

Chapter 8
Simulation and Validation

Abstract Time complexity analysis assesses the behavior of the algorithms asymptotically. However, simulation is necessary to assess differences between Privacy-Preserving Protocols (PPPs) in a realistic scenario. Many values change the behavior of the algorithms, for instance, measurements $m_{i,j}$, number of users $\tilde{i}$, number of rounds $\tilde{j}$, etc. For this reason, the parameters and dataset used should be as close to real as possible.

Keywords Simulation • Parameters • Real-world dataset • Inconsistencies • Performance • Implementation • Time

Section 8.1 presents the real-world dataset used, its inconsistencies found, and how it was sanitized. Section 8.2 presents the tools used to implement the core algorithms of the PPPs. Section 8.3 presents the parameters used in the algorithms, and Sect. 8.4 presents the simulation results that validate the performance analysis.

8.1 Dataset

A total of 6,435 meters located in Irish homes and businesses collected measurements every 30 min for one and half years. The first round refers to the consumption from 00:00:00 to 00:29:59 Greenwich Mean Time (GMT) on Tuesday, July 14, 2009. The last round refers to the consumption from 23:30:00 to 23:59:59 GMT on Friday, December 31, 2010. Thus, the dataset was composed of 25,726 rounds, namely a round for every half hour during one and a half years. The number of measurements in the raw dataset is 157,992,996, but it should be the product of the numbers of meters by rounds.

Section 8.1.1 presents the inconsistencies found in the dataset. Section 8.1.2 presents how the dataset was sanitized. Section 8.1.3 presents the dataset characteristics, i.e., the amount of information in the dataset.

© Springer International Publishing Switzerland 2017   111
F. Borges de Oliveira, *On Privacy-Preserving Protocols for Smart Metering Systems*,
DOI 10.1007/978-3-319-40718-0_8

Table 8.1 Duplicated measurements

Date	Timestamps	Power (KW)	Meter
Sat 24 Oct 2009 23:59:59 GMT	29,748	0.41	File1.txt:1208
Sat 24 Oct 2009 23:59:59 GMT	29,748	0.461	File1.txt:1208
Sat 24 Oct 2009 23:59:59 GMT	29,748	0.388	File5.txt:5221
Sat 24 Oct 2009 23:59:59 GMT	29,748	0.992	File5.txt:5221
Sun 25 Oct 2009 23:59:59 GMT	29,848	0.143	File1.txt:1208
Sun 25 Oct 2009 23:59:59 GMT	29,848	0.415	File1.txt:1208
Sun 25 Oct 2009 23:59:59 GMT	29,848	0.401	File5.txt:5221
Sun 25 Oct 2009 23:59:59 GMT	29,848	1.312	File5.txt:5221
Mon 26 Oct 2009 23:59:59 GMT	29,948	0.201	File1.txt:1208
Mon 26 Oct 2009 23:59:59 GMT	29,948	1.006	File1.txt:1208
Mon 26 Oct 2009 23:59:59 GMT	29,948	1.205	File5.txt:5221
Mon 26 Oct 2009 23:59:59 GMT	29,948	1.312	File5.txt:5221
Tue 27 Oct 2009 23:59:59 GMT	30,048	0.212	File1.txt:1208
Tue 27 Oct 2009 23:59:59 GMT	30,048	1.671	File1.txt:1208
Tue 27 Oct 2009 23:59:59 GMT	30,048	1.182	File5.txt:5221
Tue 27 Oct 2009 23:59:59 GMT	30,048	1.38	File5.txt:5221
Wed 28 Oct 2009 23:59:59 GMT	30,148	0.576	File1.txt:1208
Wed 28 Oct 2009 23:59:59 GMT	30,148	1.019	File1.txt:1208
Wed 28 Oct 2009 23:59:59 GMT	30,148	0.522	File5.txt:5221
Wed 28 Oct 2009 23:59:59 GMT	30,148	0.896	File5.txt:5221
Thu 29 Oct 2009 23:59:59 GMT	30,248	0.163	File1.txt:1208
Thu 29 Oct 2009 23:59:59 GMT	30,248	0.456	File1.txt:1208
Thu 29 Oct 2009 23:59:59 GMT	30,248	0.386	File5.txt:5221
Thu 29 Oct 2009 23:59:59 GMT	30,248	0.463	File5.txt:5221

8.1.1 Anomalies

This work detected some inconsistencies in the dataset [2]. They are the registry of unknown failures or attacks. In general, they register anomalies generated in the data collection. Specifically, the meters addressed as 1208 and 5221 sent two different measurements in the same round j. The measurement $m_{i,j}$ represents the consumption of a meter i in a round j. Therefore, for fixed value of i and j, the measurement $m_{i,j}$ has a unique fixed value. Table 8.1 shows the 24 duplicated measurements found in the raw dataset in which 12 are surely wrong.

The first three digits of a timestamp represent a date and are called date address, while the last two digits represent an hour and a minute and are called time address. Since the meters collected the measurements with intervals of 30 min, there are 48 time addresses per day, namely 1 is the interval from 00:00:00 to 00:29:59, 2 is the interval from 00:30:00 to 00:59:59, 3 is the interval from 01:00:00 to 01:29:59, etc. However, the raw dataset has 25,002 messages with time addresses bigger than 48, but none were set to zero and no entry in the dataset is negative. The great

Table 8.2 Undocumented timestamps

Quantity	Date address	Date
94	297	Sat 24 Oct 2009
12,658	298	Sun 25 Oct 2009
94	299	Mon 26 Oct 2009
94	300	Tue 27 Oct 2009
94	301	Wed 28 Oct 2009
94	302	Thu 29 Oct 2009
11,874	669	Sun 31 Oct 2010

majority of the time addresses are set with 49 and 50. Table 8.2 shows the quantity of messages with time addresses bigger than 48 grouped by the date address. In addition, Table 8.2 shows the date with respect to the date address. The dataset description does not specify these values, but 12,568 and 11,874 messages may be in the raw dataset due to daylight saving time on the last Sunday of October in 2009 and 2010, respectively. Besides daylight saving time, the raw dataset contains 20 measurements with undocumented time addresses from 49 to 50. In addition, it contains 540 messages with undocumented time addresses from 51 to 95. Among them, 90 messages have date addresses equal to 298, i.e., they were on Sun 25 Oct 2009. Note that $94 \times 5 + 90 - 20 = 540$ and $540 + 20 + 12,568 + 11,874 = 25,002$.

These inconsistencies are important to draw attention to the fact that PPPs should be able to verify whether the bills $b_i^\$$ are correct and whether the consolidated consumptions c_j are in agreement with the aggregated measurements a_j.

8.1.2 Sanitized Dataset

Since the PPPs cannot verify and cannot request new measurements, the inconsistencies were eliminated from the dataset, including the probable results from daylight savings time. For this performance analysis, such data are not relevant. However, it has a financial impact, because a total of only 263 messages with time addresses bigger than 48 reported measurements equal to zero.

The timestamps determine the rounds j. Thus, each meter should have collected 25,726 measurements. The missing measurements were filled as zero. For the privacy point of view, it is important that all meters contribute to consolidated consumption c_j. Otherwise, few meters might send their measurements. As a result, an attacker might get information of a specific pattern of power consumption. In addition, the verification processes will not work, because the counting agent cannot decide whether all messages were received. Moreover, if the meters do not report null measurements $m_{i,j}$, attackers know when there is no consumption.

The raw dataset has 1,557,479 measurements equal to zero. After the elimination of the anomalies, the number of zeros was 1,557,216. In order for each meter i to contribute in each round j, the unreported messages were filled with zeros. Thus, the sanitized dataset has 7,578,840 more zeros than the raw dataset, resulting in 9,136,056 zeros. Therefore, the sanitized dataset has 165,546,810 measurements collected by 6,435 meters in 25,726 rounds.

From now on, this chapter uses only the word dataset to refer to the sanitized dataset.

8.1.3   Dataset Characteristics

This section highlights some variances in the data set and aims to give a tangible idea of the values in the dataset, because they can change the performance of PPPs. For example, modular exponentiations of the measurements have performance depending on them [4]. Some information that influence the performance is in the previous section, for instance, fully connected Symmetric DC-Nets (SDC-Nets) depend on the number of meters in the encryption algorithm.

Figure 8.1 depicts the distribution of measurements. Figure 8.1a depicts the box plot of all measurements $m_{i,j}$ for all meters i and all rounds j. The lower measurement $m_{i,j}$ has zero value and the higher has a value of 66,815 W. The outliers are expected due to the high difference between the median given by 256 and the arithmetic mean marked as a red dot and given by approx. 673.39.

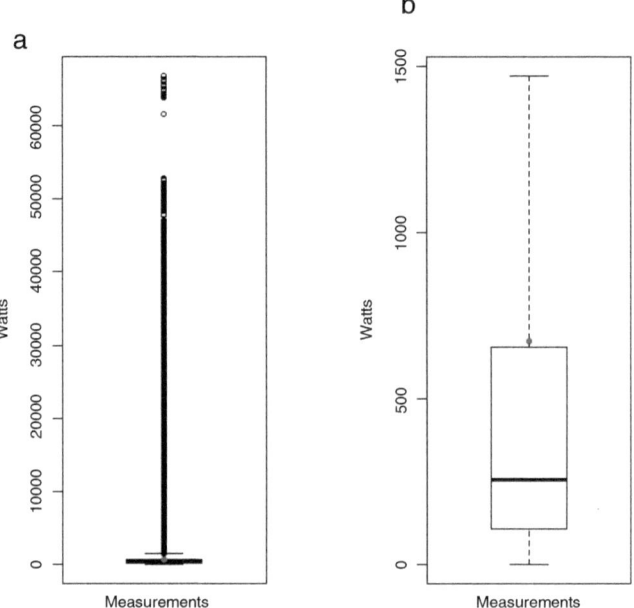

Fig. 8.1   Box plot of measurements. (**a**) Box plot with outliers. (**b**) Box plot without outliers

Figure 8.1b depicts the box plot without outliers, and we may see the difference between the median and the mean marked as a red dot. Excluding outliers, the maximum value of measurement is given by 1,469, and the minimum is still zero. The upper and lower quartiles are given by 653 and 109, respectively. Note that the mean is bigger than the upper quantile meaning outliers beyond the upper inner fences.

The biggest arithmetic mean of measurements grouped by meter is approx. 19,869.38, and the smallest is approx. 0.26. Appendix C presents Fig. C.1 that depicts the bar plot of the arithmetic mean of all measurements collected per each meter. Figure 8.2 depicts the variations of the arithmetic means in Fig. C.1. As the measurements in Fig. 8.1a, the box plot of their arithmetic mean depicted in Fig. 8.2a has many outliers. Thus, Fig. 8.2b depicts the box plot without outliers. Excluding them, its smallest mean is still the same, but its biggest is approx. 1,277.9. Its lower quartile is approx. 316.2, and its upper quartile is approx. 703.69. Its median is approx. 488.64, but the arithmetic mean of the means is approx. 673.39. Different from Fig. 8.1, the mean is behind the upper quantile in Fig. 8.2.

The distribution generated by measurements might be different by months or days of the week. Figure 8.3 depicts the mean consumption by month, while Fig. 8.4 depicts the consumption by days of the week. Figure 8.3 depicts a bar plot of the arithmetic mean of measurements grouped by year and month. We can observe that the electricity consumption decreases from the beginning of year to the middle of the

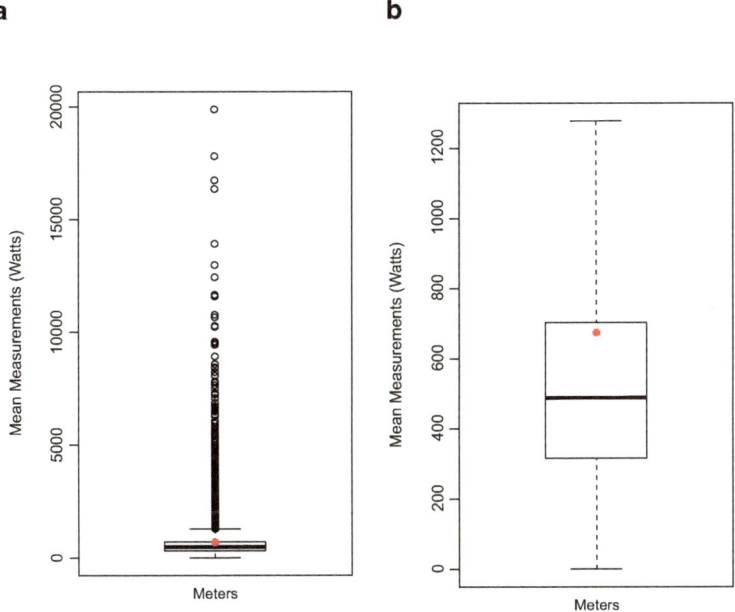

Fig. 8.2 Box plot of measurement arithmetic means grouped by meter. (**a**) Box plot with outliers. (**b**) Box plot without outliers

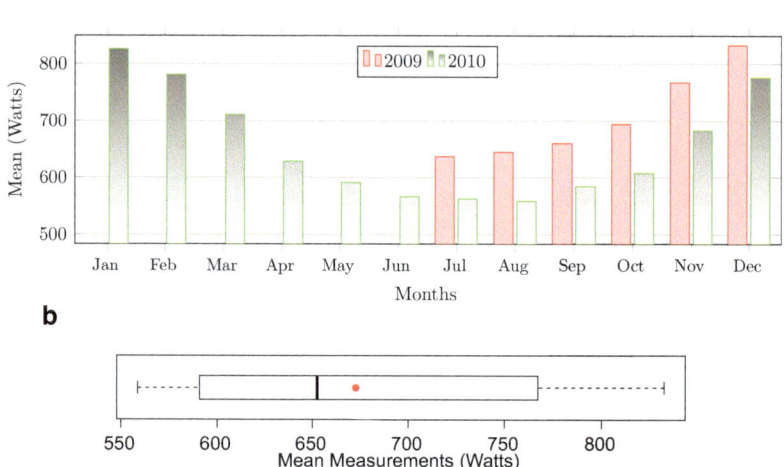

Fig. 8.3 Arithmetic mean of all measurements grouped by month. (**a**) Bar plot. (**b**) Box plot

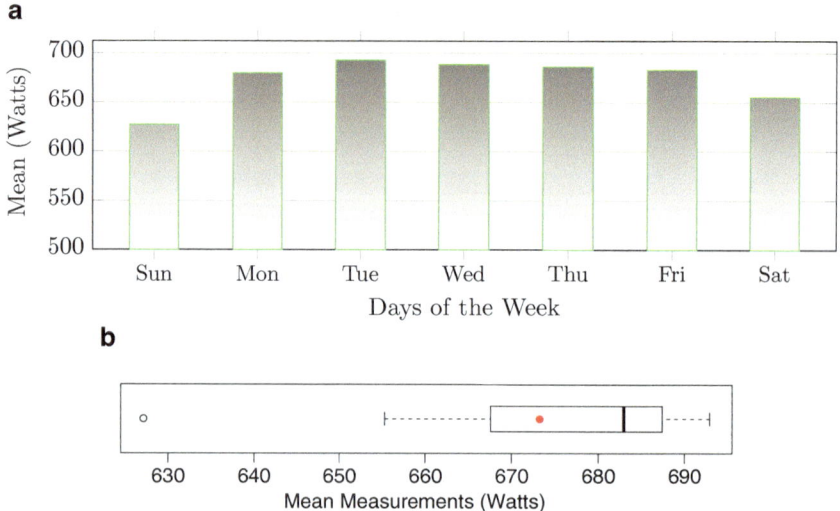

Fig. 8.4 Arithmetic mean of all measurements grouped by days of the week. (**a**) Bar plot. (**b**) Box plot

year, and it increases from the middle to the end of the year. Moreover, the reported consumption in 2009 was bigger than in 2010. The pattern in Fig. 8.3 might be correlated with the seasons, especially, with the winter.

Fig. 8.5 Number of bits used per consolidated consumption

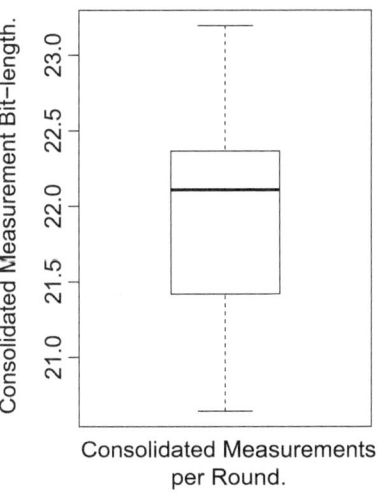

Consolidated Measurements per Round.

The minimum is approx. 558.75 and occurs in December 2009. The lower quartile given by approx. 591.04 is represented by May of 2010. The arithmetic means of consumption during June, July, and September of 2010 are between the lower quartile and minimum. The arithmetic means of consumption during July and August of 2009 and April and October of 2010 are between the lower quartile and the median given by approx. 652.49, which is close to the mean given by approx. 672.84. The arithmetic means of consumption during September and October of 2009 and March and November of 2010 are between the median and the upper quartile given by approx. 767.3 is represented by November of 2009. The arithmetic means of consumption during January, February, and December of 2010 are between the upper quartile and the maximum given by approx. 832.54 and is represented by December of 2009. Figure 8.3a depicts a bar plot of the arithmetic means grouped per months and years, and Fig. 8.3b depicts a box plot of them.

Figure 8.4 depicts the arithmetic mean of the measurements grouped by days of the week. On weekends, the consumption is smaller than on weekdays. Considering Sundays as an outlier, the minimum is approx. 655.26 and refers to Saturdays. The median is approx. 682.94 and refers to Fridays, and the mean is approx. 673.28. The lower quartile is approx. 667.56 and smaller than Mondays. The upper quartile is approx. 687.39, bigger than Thursdays and smaller than Wednesdays. The maximum is approx. 692.96 and occurs on Tuesdays. Figures 8.4a, b depict a bar plot and a box plot of the means grouped by days of the week, respectively.

In general, the size of the measurements $m_{i,j}$ has a small influence on the processing time of each encryption algorithm and has a smaller influence on the processing time of the aggregation and decryption algorithms. Specifically for PPP3, the size of the consolidated consumption c_j has a strong influence in the search for its value. The processing time of PPP3 may define if it can be used as encryption–decryption scheme or only as a commitment scheme. Figure 8.5 depicts the box plot of the number of bits used per consolidated consumption c_j.

8.2   Implementation of the Core Algorithms

The source codes of the cryptographic algorithms—encryption, aggregation, and decryption—were implemented for the PPPs, namely, PPP1, PPP2, PPP3, LOOP [3], EPPP4SMS [2], and Paillier [5] that is used in several protocols. The decryption of PPP3 was implemented as an open function for commitments. The algorithms were implemented without optimizations, e.g., without precomputation. Thus, the simulation used the modular exponentiation given by Algorithm 16 instead of modular multi-exponentiation given by Algorithm 17.

The algorithms were written in the C programming language and compiled with GCC version 4.6.1 for GNU/Linux with the Ubuntu distribution. They were linked with the GNU Multiple Precision Arithmetic Library (GMP), the Open Multi Processing (OpenMP), and the open source toolkit for SSL/TLS (OpenSSL). GMP was used to manipulate big numbers. OpenMP was used to parallelize the encryption. OpenSSL was used to run the hash function.

The simulation ran in a machine with Intel® Xeon®, CPU E5-2660 of 2.20GHz, 32 recognized cores, and 63 Gigabytes of shared memory.

8.3   Simulation Parameters

This simulation ran with the same hash function for all PPPs. Previous simulations ran with different hash functions to give an advantage to SDC-Net, but this strategy is not enough to make a fully connected SDC-Net faster than other PPPs [1]. Simulations of SDC-Net, Pascal, etc. with the same dataset and with different hash functions were previously done [1, 2]. The hash function chosen for this book simulation was SHA256.

Asymmetric DC-Nets (ADC-Nets) are represented by Low Overhead protocol (LOP) [3], which only uses 32 bits of the hash function; cf. Sect. 3.2.2. Similarly, PPP3 only uses 160 bits of the hash function. Accordingly, the keys for LOP and PPP3 have 32 and 160 bits, respectively. PPP1 has the same key length as LOP has, i.e., 32 bits. PPP2 has 160 bit of key length. EPPP4SMS [2] has two keys of 160 bits.

Only for comparison, the elliptic curve $\Omega(\mathbb{Z}_p)$ selected was the P-192 standardized by National Institute of Standards and Technology (NIST) in its publication FIPS 186-2. Appendix B describes the P-192. Given Ω and the hash function H, then H_{Ω} is defined in Eq. (6.10). The function H_{Ω} searches for the first x bigger than $H(j)$ that satisfies

$$x^{\frac{p+1}{2}} \equiv 1 \mod p,$$

and it determines

$$y = x^{\frac{p+1}{4}} \mod p.$$

In addition, it computes a scalar multiplication of the point (x, y) by n as defined in Appendix B. If the result is different from the point of infinity, the program returns an error message. Such computation ensures that the function H_Ω returns a safe point, i.e., a point with high order. Nevertheless, this curve has $h = 1$—cf. Appendix B—and such computation is not necessary for $h = 1$.

Paillier, EPPP4SMS, and PPP3 require the product of two primes with 512 bits each. In this case, their product has 1,024 bits. They are chosen by the cryptographically secure pseudorandom number generator (CSPRNG) implemented in the GMP. Using the same generator, the key size is determined by the smaller security level in Table 6.1, which also gives us the smaller difference of processing time between the protocols.

To speed up the simulation, the encryption algorithms were parallelized in 30 threads.

8.4 Simulation Results

Before presenting the processing time collected per each round of each algorithm, let us find whether PPP2 can be used as an encryption–decryption function. Figure 8.6 depicts the processing time for the number of threads to search for the consolidated consumption c_j given the encrypted consolidated consumption in PPP2.

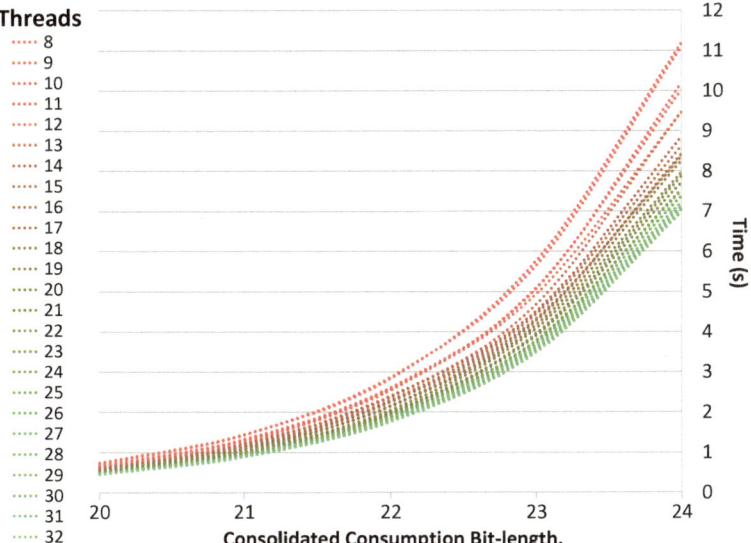

Fig. 8.6 Parallel brute force search for a c_j in PPP2

On the one hand, the supplier runs the most computationally intensive part of PPP2. In contrast to the meters, the supplier has enough computational power to search for consolidated consumptions c_j. On the other hand, the meters need to run a light computation over an elliptic curve. With this scenario, we may consider the PPP2 as an encryption–decryption scheme instead of only a commitment scheme. Note that the supplier can search with much more than 32 threads. In addition, the supplier can use the aggregated measurement a_j to speed up the search for the consolidated consumption c_j. Considering the number of bits necessary for the search is approx. 22—cf. Fig. 8.5—then the search spends less than 2 s for PPP2 to find a consolidated consumption c_j, cf. Fig. 8.6. The time can be significantly reduced with information from the phasor measurement units (PMUs), i.e., the aggregated measurements a_j.

8.4.1 Encryption Algorithms

Figure 8.7 depicts the box plot of the time observed per round j for the running time of the encryption algorithms.

As expected, the PPP1 is the fastest with a processing time of approx. 16.9 ms per round. LOP has the same key length as PPP1, but it is the most expensive with a mean of more than 10 s per meter. PPP1 computes $\tilde{\imath}$ hash functions per round, where $\tilde{\imath}$ is the number of users, while LOP computes $\tilde{\imath}^2$ hash functions. Surprisingly, LOP ran approx. 10 times faster than expected, i.e., approx. $\tilde{\imath} \times 0.0169 = 6435 \times 0.0169 = 108.7515$ instead of 10.4582 s. Probably, LOP was fast due to automatic optimizations for parallelization and hardware. Although PPP2 and PPP3 have the same key length, PPP3 is more than twice as fast as PPP2 on average. As expected, PPP3 has processing time close to two times faster than EPPP4SMS. Paillier ran in approx. 1 s per round, i.e., 10 times faster than LOP.

Table 8.3 shows the result of the statistical analysis used to plot Fig. 8.7. The maximum and minimum in Table 8.3 were computed without the outliers.

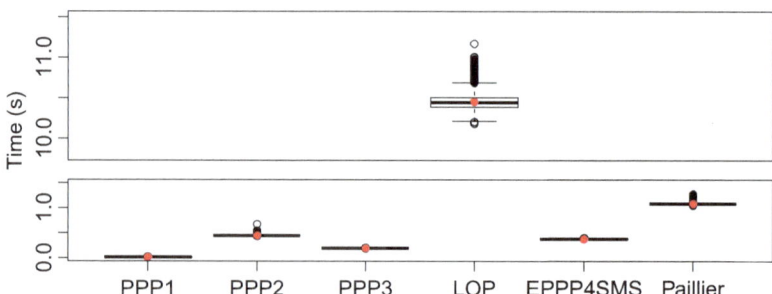

Fig. 8.7 Box plot of encryption algorithms

Table 8.3 Time observed for encryption algorithms in seconds

	PPP1	PPP2	PPP3	LOP	EPPP4SMS	Paillier
Min.	0.0117	0.4390	0.1895	10.2162	0.3790	1.0796
Lower quartile	0.0155	0.4467	0.1941	10.3937	0.3835	1.0843
Mean	0.0169	0.4510	0.1957	10.4582	0.3866	1.0876
Median	0.0168	0.4490	0.1955	10.4482	0.3877	1.0858
Upper quartile	0.0180	0.4518	0.1972	10.5127	0.3887	1.0875
Max.	0.0219	0.4595	0.2017	10.6912	0.3964	1.0922

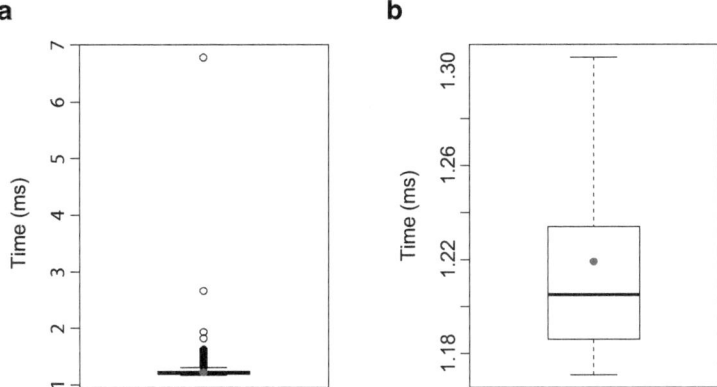

Fig. 8.8 Box plot of the hash for elliptic curves H_Ω. (**a**) Box plot with outliers. (**b**) Box plot without outliers

The collected time for EPPP4SMS does not consider the hash function for elliptic curves H_Ω given in Eq. (6.10). The computation of H_Ω is not efficient. PPP3 is feasible only if the meters distributed the task and compute only one hash for all meters. Figure 8.8 depicts the processing time to compute H_Ω. Figure 8.8a depicts all collected times, while Fig. 8.8b depicts the observed processing time excluding the outliers. The minimum time is approx. 1.171 ms, the lower quartile is approx. 1.186 ms, the median is approx. 1.205 ms and is relatively far from the mean that is approx. 1.219 ms, the upper quartile is approx. 1.234 ms, and the maximum observed time is approx. 1.306 ms.

8.4.2   Aggregation Algorithms

Figure 8.9 depicts the box plot of the time observed for the aggregation algorithms. Note that the spot for LOP is empty because it is the only protocol that does not need an aggregation algorithm. Precisely, LOP aggregates the encrypted measurements $\mathfrak{M}_{i,j}$, but its aggregation cannot be split from the decryption algorithm. Thus, all the cost for aggregations is in the decryption algorithm.

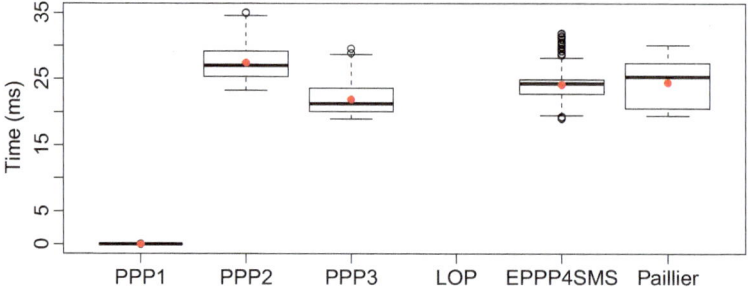

Fig. 8.9 Box plot of aggregation algorithms

Table 8.4 Time observed for aggregation algorithms in milliseconds

	PPP1	PPP2	PPP3	LOP	EPPP4SMS	Paillier
Min.	0.0230	23.2690	18.9830	NA	19.4940	19.4670
Lower quartile	0.0240	25.3330	20.0950	NA	22.7190	20.5480
Mean	0.0280	27.4186	21.9030	NA	24.1553	24.4537
Median	0.0240	27.0340	21.2580	NA	24.3070	25.3160
Upper quartile	0.0250	29.1840	23.6170	NA	24.9070	27.3920
Max.	0.0260	34.6080	28.6980	NA	28.1410	30.0810

Excluding LOP, PPP1 is the fastest to aggregate with a mean of approx. 0.028 ms. As expected, PPP3, EPPP4SMS, and Paillier have equivalent running time with mean of 21.9030, 24.1553, and 24.4537 ms, respectively. They have the same parameter size and use the same code in C language as follows:

```
start_clock();
mpz_set_ui(st,1); // st = 1
for(i=0;i<6435;i++) {
        // commands equivalent to st = st*C[i] % n^2
        mpz_mul(s, st, C[i]); // multiplication
        mpz_mod(st, s, n2); // modulo
}
end_clock(prod);
```

However, the primes, the keys, etc. are generated pseudo-randomly.

PPP2 has a different code and works over an elliptic curve resulting in a running time of approx. 27.4186 ms. Although Elliptic Curve Discrete Logarithm Problem (ECDLP) works with smaller numbers than Integer Factorization Problem (IFP), the point addition requires subtractions, inversion, squaring, and multiplication over a field.

Table 8.4 shows the data of running time used to plot Fig. 8.9. Since LOP has no aggregation algorithm, its column in Table 8.4 contains not applicable (NA). For LOP, the aggregation happens together with the decryption. The Max. and Min. are given without considering the outliers.

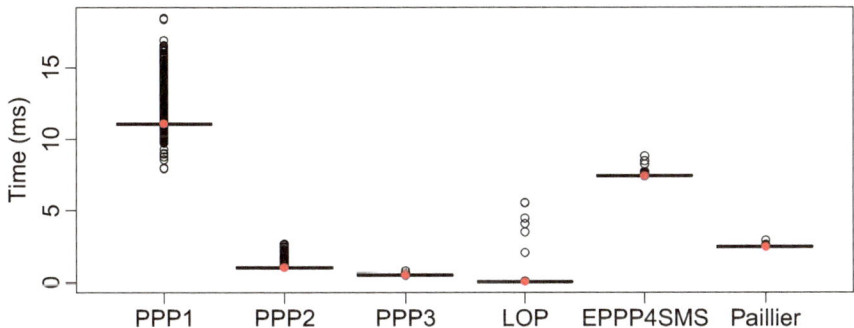

Fig. 8.10 Box plot of decryption algorithms

Table 8.5 Time observed for decryption algorithms in milliseconds

	PPP1	PPP2	PPP3	LOP	EPPP4SMS	Paillier
Min.	11.0000	0.9750	0.4760	0.0170	7.4070	2.4600
Lower quartile	11.0500	1.0060	0.4880	0.0210	7.4190	2.4650
Mean	11.0955	1.0232	0.4922	0.0338	7.4242	2.4686
Median	11.0660	1.0140	0.4920	0.0300	7.4220	2.4680
Upper quartile	11.0840	1.0380	0.4960	0.0430	7.4270	2.4710
Max.	11.1340	1.0860	0.5080	0.0750	7.4390	2.4800

8.4.3   Decryption Algorithms

Figure 8.10 depicts the box plot of the observed running time in milliseconds for the decryption algorithms per round.

PPP1 is approx. 11 times slower than PPP, which is approx. 2 times slower than PPP3, which is approx. 14 times slower than LOP. Excluding the outliers, LOP is the fastest to decrypt. PPP2 is approx. 7 times faster than EPPP4SMS and 2.5 times faster than Paillier. Considering that the times are in milliseconds and that decryption algorithms run on the supplier side, the differences of running time between the algorithms are not as significant as for the encryption algorithms, which run on the meter side and their running time is presented in seconds. Decryption algorithms ran faster than aggregation algorithms, excluding PPP1 and LOP.

Table 8.5 contains the values used to plot Fig. 8.10. The outliers are not considered to be the maximum and the minimum value.

Similar to the collected times for encryption algorithms, the collected time for decryption or opening the commitment in PPP2 does not consider the hash function H_Ω given in Eq. (6.10).

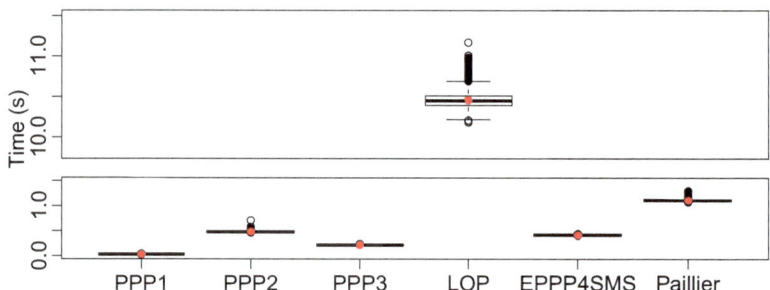

Fig. 8.11 Box plot of the overall time observed per round j

Table 8.6 Overall time observed per round j in seconds

	PPP1	PPP2	PPP3	LOP	EPPP4SMS	Paillier
Min.	0.0226	0.4668	0.2099	10.2163	0.4121	1.0957
Lower quartile	0.0266	0.4757	0.2149	10.3938	0.4170	1.1080
Mean	0.0281	0.4807	0.2181	10.4582	0.4182	1.1145
Median	0.0279	0.4791	0.2176	10.4482	0.4196	1.1149
Upper quartile	0.0293	0.4816	0.2213	10.5128	0.4202	1.1163
Max.	0.0334	0.4905	0.2299	10.6913	0.4249	1.1286

8.4.4   Overall Performance

In summary, the encryption algorithms have higher impact in the overall performance than their respective aggregation and decryption algorithms. Figure 8.11 depicts the box plot of the overall time observed per round j.

Figure 8.11 depicts the total processing time per round j for each protocol, i.e., the time observed for encryption, aggregation, and decryption algorithms. In the case of PPP3, the time for computing one hash function H_Ω is also included. As expected, Fig. 8.11 is similar to Fig. 8.7, which depicts the time for encryption algorithms. Table 8.6 shows the values used to plot Fig. 8.11.

Metaphorically, the meters are runners ready to sprint. With this analogy, we can compare the time necessary for meters to process a round with different PPPs. Each meter on running track has the same hardware but different protocols. The software determines the winner. Figure 8.12 depicts the exact moment in which PPP1 wins the competition. Indeed, PPP1 is fastest and much faster than the second classified, i.e., PPP3.

Note that PPP3 is approx. two times faster than PPP2 and EPPP4SMS. PPP1 uses approx. half of its time in the encryption algorithm, and the other half in the decryption algorithm. The time for its aggregation algorithms is insignificant in comparison. Although PPP1 has the biggest difference between encryption and

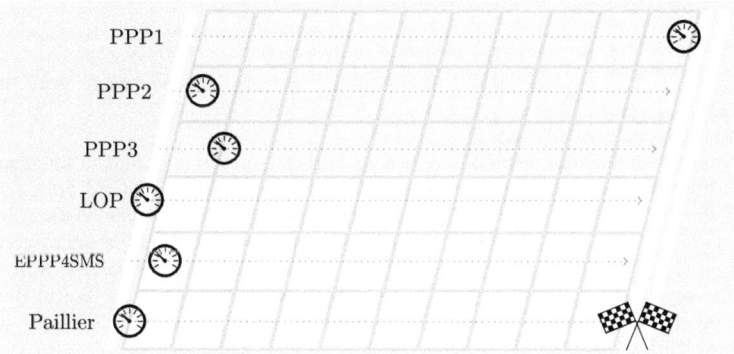

Fig. 8.12 Meter race

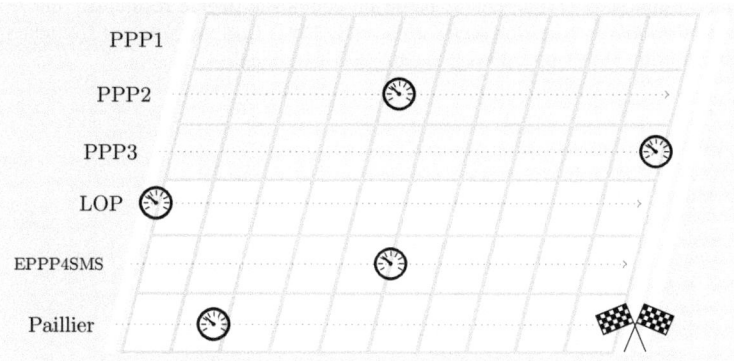

Fig. 8.13 Meter race without PPP1

overall performance, it is still the fastest overall, but meters and suppliers need to run PPP1 with PPP2 to have features of PPP3. Therefore, PPP3 outperforms the others.

The meter metaphor can be used again for us to compare better the performance of the protocols excluding PPP1, which has an isolated performance. Figure 8.13 depicts the moment when PPP3 wins the competition.

The meter racing shows that PPP1 is really the fastest, and PPP3 heavily outperforms the other PPPs, because PPP3 has several interesting properties as described in Chaps. 6 and 7.

Acknowledgements For the dataset, I thank the Irish Social Science Data Archive and the Commission for Energy Regulation (CER), Electricity Customer Behaviour Trial, issued by The Research Perspective Ltd. on 12–03–2012. I thank the Brazilian National Laboratory for Scientific Computing (LNCC in Portuguese) for the infrastructure.

References

1. F. Borges, L.A. Martucci, iKUP keeps users' privacy in the smart grid, in: *2014 IEEE Conference on Communications and Network Security (CNS)* (2014), pp. 310–318; doi:10.1109/CNS.2014.6997499
2. F. Borges, M. Mühlhäuser, EPPP4SMS: efficient privacy-preserving protocol for smart metering systems and its simulation using real-world data. IEEE Trans. Smart Grid **5**(6), 2701–2708 (2014); doi:10.1109/TSG.2014.2336265, url:http://dx.doi.org/10.1109/TSG.2014.2336265
3. K. Kursawe, G. Danezis, M. Kohlweiss, Privacy-friendly aggregation for the smart-grid, in *Proceedings of the Privacy Enhancing Technologies: 11th International Symposium, Waterloo, ON, Canada, July 27–29, 2011*, ed. by S. Fischer-Hübner, N. Hopper (Springer, Berlin/Heidelberg, 2011), pp. 175–191; isbn:978-3-642-22263-4, doi:10.1007/978-3-642-22263-4_10, url:http://dx.doi.org/10.1007/978-3-642-22263-4_10
4. P. Lara et al., Parallel modular exponentiation using load balancing without precomputation. J. Comput. Syst. Sci. **78**(2), 575–582 (2012); issn:0022-0000, doi:10.1016/j.jcss.2011.07.002, url:http://dx.doi.org/10.1016/j.jcss.2011.07.002
5. P. Paillier, Public-key cryptosystems based on composite degree residuosity classes, in *Advances in Cryptology - EUROCRYPT 1999*, vol. 1592. Lecture Notes in Computer Science (Springer, Berlin, 1999), pp. 223–238; isbn:978-3-540-65889-4

Chapter 9
Concluding Remarks

Abstract After three chapters contextualizing this research followed by five chapters advancing the state of the art, this book recapitulates itself, summarizes the main results, presents an outlook, and synthesizes its importance.

Keywords Synthesis • Recapitulation • Summary • Retrospective • Perspectives • Remarks

9.1 Recapitulation

On the one hand, smart grids can provide numerous benefits to society, which, as a result, can be wealthier with energy that is more sustainable and eco-friendly. On the other hand, they can also enable surveillance and manipulation of society and private citizens. The solution to this dilemma is found in Privacy-Preserving Protocols (PPPs), cf. Chap. 1. Despite the privacy problem, smart grids have been deployed around the world in parallel with research in security and privacy, cf. Chap. 2. The majority of the related work has addressed the consolidated consumption c_j, but smart grids have other primordial requirements, cf. Chap. 3. Addressing the bill $b_i^\$$ and consolidated consumption c_j, any PPP can be attacked by means of algebra and probability, cf. Chap. 5. To compute safe aggregations, the protocols need to be efficient. Four interesting PPPs are selected. Besides these solutions, the concept of Asymmetric DC-Nets (ADC-Nets) is introduced from the concept of Symmetric DC-Nets (SDC-Nets), cf. Chap. 6. The former needs only a key per smart meter. Thus, the number of keys grows linearly. The latter needs $\tilde{\imath}$ keys per smart meter. Thus, the number of keys has quadratic growth with respect to the number of smart meters $\tilde{\imath}$. Each protocol has its advantages and disadvantages. Overall, ADC-Nets are the most satisfactory and can have desirable properties, including the benefits from SDC-Nets and additive homomorphic encryption primitives (AHEPs), cf. Chap. 7. The verification capability is indispensable, and the real-world dataset used reaffirms the necessity of verification with its inconsistencies. The simulation agrees with theoretical results and shows that the PPP3—an ADC-Net—outperforms the others, cf. Chap. 8.

© Springer International Publishing Switzerland 2017   127
F. Borges de Oliveira, *On Privacy-Preserving Protocols for Smart Metering Systems*,
DOI 10.1007/978-3-319-40718-0_9

9.2   Main Results

This book contains a variety of results regarding PPPs for smart metering systems, including the improvement of published PPPs. Such development culminated in the concept of ADC-Nets, which are generalizations of AHEPs. In summary:

- Remote and frequent measurements are important.
- Suppliers need only the encrypted measurements.
- PPPs for smart metering systems should provide:

  - Consolidated consumptions c_j;
  - Bill $b_i^\$$;
  - Verification (auditability);
  - Efficiency.

- PPPs only preserve privacy for large aggregations.
- Optimal aggregation when $\tilde{i} = c_j$ and $\tilde{j} = b_i$.
- PPP1 is conjectured to be the fastest on the meter side.
- The speed advantage of PPP2 and PPP3 in comparison with other PPPs is exponentially higher with increasing key size.
- The concept of ADC-Nets is introduced.
- PPP4 has different properties due to quantum mechanics.
- The analytical comparison shows that ADC-Nets are supreme.
- Inconsistencies lead to verifications.
- The simulation validates the theoretical results.

9.3   Outlook

Besides the relation between fully homomorphic encryption and ADC-Nets, this book provides new perspectives for research in areas such as algorithms, cryptography, security, privacy, and smart grid. As mentioned, many application scenarios may use the results of this book, cf. Sect. 6.4.1.3. In summary, new research can be done based on this book as follows.

Chapter 4 presents economic reasons for suppliers to compute consolidated consumptions. New reasons can be identified. Perhaps, some of them might require more than the minimum requirements in Sect. 4.2. Fair distribution provides a challenge for PPPs with untrusted smart meters in Sect. 4.1.3. The description of the application scenario and the identification of the requirements are crucial to the development of new technologies and PPPs.

Chapter 5 quantifies the risks as a function of several variables. However, it does not present a minimum number of smart meters to generate consolidated consumptions with a satisfactory level of privacy. This problem can be assessed with experimental approaches. Nevertheless, different populations might have different minimum. New algebraic and probabilistic relations can be found.

Chapter 6 improves four published protocols. The search for more efficient PPPs is
fundamental. However, it is paramount that the new PPPs can satisfy the minimum requirements presented in Sect. 4.2 and can enforce privacy as ADC-Nets
can. PPP1 and PPP3 have exponential gain in performance. Thus, it is difficult
to find algorithms with better complexity. However, they can be improved with
low-level implementation. ADC-Nets can be explored in theory and application,
i.e., studying its relations with fully homomorphic encryption and using it for
different proposals, e.g., e-voting, reputation systems, trust, sensor networks,
multi-party computation, e-cash, mobile sensing, image processing, etc.

Chapter 7 compares different PPPs. New PPPs should be at least compared with
PPP3, i.e., ADC-Nets, which can be compared with PPPs for other applications.
Systematic reviews boost new developments.

Chapter 8 validates the performance of the selected PPPs by means of processing
time. New simulations can include communication costs, can be done in real
smart meters, can use a dataset with all the customers of a supplier, etc. An
ADC-Net can enforce privacy for subsets of customers to avoid the need of
setting up the whole PPP again if a smart meter fails. The subsets might also
have intersections. Determining the optimal configuration is a challenge, which
depends on several variables and may be determined with simulation.

9.4 Final Remarks

Smart grids can leverage society's resources and enhance its economy. However,
security and privacy in smart grids are fundamental for every nation. Therefore,
PPPs can speed the proper development of society. Many techniques can be used
to keep customers' privacy secure. Nonetheless, few techniques enforce privacy in
their PPPs. Without the correct enforcement by means of cryptography, the right to
security and privacy might be violated eventually.

Appendix A
Algorithms

The variables of the algorithms in this appendix are not related to predefined variables, i.e., they are local variables.

Algorithm 16: Modular exponentiation

Input: Integers b, e, n s.t. $e = \sum_{\iota=1}^{l} 2^{\iota-1} e_\iota$, where l is the bit length of e and $e_\iota \in \{0, 1\}$.

Output: $b^e \mod n$.

1   $a \leftarrow 1$
2   **for** $\iota = l$ **to** 1 **by** -1 **do**
3     $a \leftarrow a^2 \mod n$
4   **if** $e_i = 1$ **then**
5       $a \leftarrow ab \mod n$

6   **return** a

Algorithm 17: Modular multi-exponentiation

Input: Integers b_i, e_i, m, n s.t. $e_i = \sum_{j=1}^{l_i} 2^{j-1} e_{ij}$, where l_i is the bit length of e_i and $e_{ij} \in \{0, 1\}$.

Output: $\prod_{i=1}^{n} b_i^{e_i} \mod m$.

1   $L \leftarrow \lceil \max(\log_2 e_1, \ldots, \log_2 e_n) \rceil$
2   $a \leftarrow 1$
3   **for** $j = L$ **to** 1 **by** -1 **do**
4     $a \leftarrow a^2 \mod m$
5   **for** $i = 1$ **to** n **do**
6   **if** $e_{ij} = 1$ **then**
7         $a \leftarrow ab_i \mod m$

8   **return** a

© Springer International Publishing Switzerland 2017   131
F. Borges de Oliveira, *On Privacy-Preserving Protocols for Smart Metering Systems*,
DOI 10.1007/978-3-319-40718-0

Algorithm 18: Finding equivalent key lengths

Output: Return a table T with the key lengths in Table 6.1.

1 **foreach** $i \in \{80,\ 112,\ 128,\ 192,\ 256\}$ **do**
2       $m \leftarrow 2^i$
3       $T \leftarrow \{\}$
4       $r \leftarrow \text{solve}\left(\sqrt{(1/2) \cdot \pi \cdot o} = m, o\right)$
5       $T_{i,\text{DLP}} \leftarrow \lceil \log_2(r) \rceil$
6       $r \leftarrow \text{solve}\left(\exp\left((64/9)^{1/3} \cdot \ln(n)^{1/3} \cdot \ln(\ln(n))^{2/3}\right) = m, n\right)$
7       $T_{i,\text{GNFS}} \leftarrow \lceil \log_2(r) \rceil$
8       $r \leftarrow \text{solve}\left(\exp\left((32/9)^{1/3} \cdot \ln(n)^{1/3} \cdot \ln(\ln(n))^{2/3}\right) = m, n\right)$
9       $T_{i,\text{SNFS}} \leftarrow \lceil \log_2(r) \rceil$
10 **return** T

Appendix B
Parameters for ECC

The elliptic curve P-192 is over $\mathbb{Z}_p$ with $p = 2^{192} - 2^{64} - 1$, and its points are defined by

$$\Omega : y^2 \equiv x^3 - 3x + b \pmod{p},$$

where

$b =$ 0x 64210519 E59C80E7 0FA7E9AB 72243049 FEB8DEEC C146B9B1.

The base point $P = (x, y)$ recommended has

$x =$ 0x 188DA80E B03090F6 7CBF20EB 43A18800 F4FF0AFD 82FF1012,

and

$y =$ 0x 07192B95 FFC8DA78 631011ED 6B24CDD5 73F977A1 1E794811.

Hence, the order of P is

$n =$ 0x FFFFFFFF FFFFFFFF FFFFFFFF 99DEF836 146BC9B1 B4D22831.

Thus, the cofactor is

$$h = \frac{|\Omega(\mathbb{Z}_p)|}{n} = 1.$$

This curve is standardized by the National Institute of Standards and Technology (NIST) in its publication FIPS 186-2.

© Springer International Publishing Switzerland 2017 133
F. Borges de Oliveira, *On Privacy-Preserving Protocols for Smart Metering Systems*,
DOI 10.1007/978-3-319-40718-0

Appendix C
Mean Measurement by Meter

See Fig. C.1.

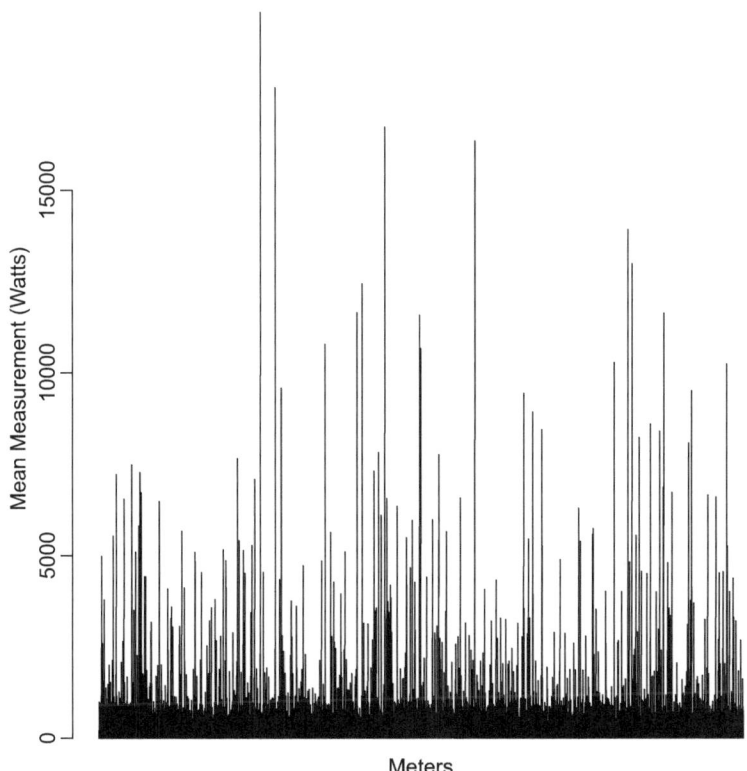

Fig. C.1 Bar plot of the mean measurement by user's meter

© Springer International Publishing Switzerland 2017
F. Borges de Oliveira, *On Privacy-Preserving Protocols for Smart Metering Systems*,
DOI 10.1007/978-3-319-40718-0

Glossary

abelian group synonym for a commutative group $(G, \circledast)$. xxvi, 67

attacker depending on the context, a customer, an employee, an adversary, a competitor, an opponent, a cryptanalyst, an enemy, or anyone who wants to exploit system vulnerabilities. 3, 7, 10, 19–22, 26, 27, 30, 45, 46, 49, 50, 53–55, 59, 84, 90, 91, 93, 97, 103, 104, 113, 137

commodity resources provided through a supply network, exhibiting the characteristics of a flow, e.g., electricity, water, gas, and heating or cooling of liquids and gases. 5, 16, 17, 41, 43–45, 62, 63, 138

counting agent an abstraction of intended recipient, which might be an supplier. xvi, 61–66, 71–78, 81, 84–91, 93–97, 102–108, 113

meter short form of smart meter. xvi, xxvi, xxvii, 104, 115, 120, 135

one-way function if it exists, it is a function that can be computed in polynomial time, but not its inverse. ix, xxvi, 8, 31, 45, 64, 65, 71, 81, 82, 85, 90, 91, 103, 107

post-quantum cryptography cryptographic algorithms that run in classical computers but are resistant against attackers with quantum computers. 7, 97

processing time time necessary to process a task, in particular, simulation data are presenting wall-clock time. xvii, 7, 8, 10, 11, 27, 44, 66, 78, 80, 83, 91, 92, 97, 106, 107, 117, 119–121, 124, 129

quantum cryptography cryptographic algorithms that need properties observed in quantum mechanics. 7, 9, 10, 61, 92, 97, 102

smart grid a network of people, computers, and sensors in a public infrastructure that monitors and manages the usage of commodities. vii–ix, xv, 3, 5–7, 9–11, 13–19, 21, 22, 25–28, 39, 40, 43, 44, 61, 79, 80, 84, 92, 101, 104, 127–129

smart meter a computerized sensor that measures the consumption of a commodity for a customer. vii–ix, 3, 5–11, 13, 15–17, 33, 39, 49, 54, 61, 78, 84, 127–129, 138

© Springer International Publishing Switzerland 2017   137
F. Borges de Oliveira, *On Privacy-Preserving Protocols for Smart Metering Systems*,
DOI 10.1007/978-3-319-40718-0

supplier a public utility that provides a commodity, e.g., energy supplier. vii–ix, xv, 3, 5–8, 10, 13, 16–19, 21, 22, 25–30, 32–34, 39–46, 49, 56, 61–63, 73, 87, 104, 120, 123, 125, 128, 129, 137

user an abstraction of a message sender, which might be a customer with a smart meter. xvi, xix, xxvi, xxvii, 3, 17, 30, 53, 61–67, 71–78, 80–91, 93–97, 103–108, 111, 120, 135

Index

A

ADC-Nets. *See* Asymmetric DC-nets (ADC-Nets)

Additive homomorphic encryption primitives (AHEPs), 7, 9, 20, 26, 101, 127, 128
  anonymization, 28–30
  performance, 105–107
  PPP1, 8
  privacy, 103
  requirements, 104
  vs. SDC-Nets and ADC-Nets, 107–108
  security, 102
  verification capabilities, 105

Advanced metering infrastructure (AMI), 13, 39

Aggregation, 7, 10
  AHEPs, 29–30
  billed consumption and consolidated consumption
    binomial, 50
    combination, 50
    possibilities in relation to sum and total, 50, 51
    probabilistic properties, 56–59
    Stirling's formula, approximation, 50, 52–53
    system of linear equations, 53–56
  data, 21
  PPP1, 8

AHEPs. *See* Additive homomorphic encryption primitives (AHEPs)

Anonymization
  cryptographic protocols for
    billing verification, 33–34
    DC-Net, 30–32
  homomorphic encryption, 28–30
  Pedersen Commitments, 32–33
  Shamir Secret Sharing, 28
  via pseudonymous, 26

Anti-collusion, 102, 103

Asymmetric DC-nets (ADC-Nets), 7, 9, 10, 30, 101, 127–129
  performance, 105–107
  PPP3, 8, 10
    aggregated measurement verification, 87, 88
    aggregation and decryption, 86
    AHEPs, 83, 84
    attacker model, 84–85
    billing verification, 89–90
    consolidated consumption, 82
    consolidated monetary value, 86
    deceptive user, detection of, 87–89
    Paillier encryption and decryption functions, 82, 83
    performance analysis, 91–92
    privacy analysis, 90–91
    probabilistic encryption, 83
    properties, 80–82
    security analysis, 90
    signed and encrypted measurement, 85
    unfriendly individual measurement verification, 87
  privacy, 103
  requirements, 104
  vs. SDC-Nets and AHEPs, 107–108
  security, 102
  simulation, 118
  verification capabilities, 105

Asymptotic complexity analysis, 105

© Springer International Publishing Switzerland 2017
F. Borges de Oliveira, *On Privacy-Preserving Protocols for Smart Metering Systems*,
DOI 10.1007/978-3-319-40718-0